# REFORM IN SCHOOL MATHEMATICS AND AUTHENTIC ASSESSMENT

SUNY Series
Reform in Mathematics Education
Judith Sowder, editor

# REFORM IN SCHOOL MATHEMATICS AND AUTHENTIC ASSESSMENT

THOMAS A. ROMBERG, Editor

State University of New York Press

The preparation of this book was supported by the Office for Educational Research and Improvement, United States Department of Education (Grant Number R117G10002) and the Wisconsin Center for Education Research, University of Wisconsin-Madison. The opinions expressed in this publication do not necessarily reflect the views of the Office of Educational Research and Improvement or the Wisconsin Center for Education Research.

Published by
State University of New York Press, Albany

© 1995 State University of New York



Printed in the United States of America



For information, address the State University of New York Press,
State University Plaza, Albany, NY 12246

Production by Bernadine Dawes • Marketing by Dana Yanulavich

**Library of Congress Cataloging-in-Publication Data**

Reform in school mathematics and authentic assessment / edited by
    Thomas A. Romberg.
        p.    cm. — (SUNY Series, reform in mathematics education)
    Includes bibliographical references and index.
    ISBN 0-7914-2161-9. — ISBN 0-7914-2162-7 (pbk.)
    1. Mathematical ability — Testing.    I. Romberg, Thomas A.
    II. Series.
    QA13.r437    1995
    510'.76—dc20

1 2 3 4 5 6 7 8 9 10

# CONTENTS


Preface       vii

1 ❖ Issues Related to the Development of an Authentic
Assessment System for School Mathematics
THOMAS A. ROMBERG AND LINDA D. WILSON       1

2 ❖ A Framework for Authentic Assessment in Mathematics
SUSANNE P. LAJOIE       19

3 ❖ Sources of Assessment Information for Instructional
Guidance in Mathematics
EDWARD A. SILVER AND PATRICIA ANN KENNEY       38

4 ❖ Assessment: No Change without Problems
JAN DE LANGE       87

5 ❖ The Invalidity of Standardized Testing for
Measuring Mathematics Achievement
ROBERT E. STAKE       173

6 ❖ Assessment Nets: An Alternative Approach to
Assessment in Mathematics Achievement
MARK WILSON       236

7 ❖ Connecting Visions of Authentic Assessment to
the Realities of Educational Practice
M. ELIZABETH GRAUE       260

Contributors       277

Index       279

# PREFACE

This volume is concerned with the alignment between the way the mathematical performance of students is assessed and the reform agenda in school mathematics. The central feature of the current reform efforts involves an epistomological shift from the mastery of a set of concepts and procedures to mathematical power. The term *mathematical power* means "an individual's abilities to explore, conjecture, and reason logically, as well as the ability to use a variety of mathematical methods effectively to solve nonroutine problems. This notion is based on recognition of mathematics as more than a collection of concepts and skills to be mastered; it includes methods of investigating and reasoning, means of communication, and notions of context. In addition, for each individual, mathematical power involves the development of personal self-confidence" (National Council of Teachers of Mathematics, 1989, p. 5).

The term *authentic assessment* has been chosen to convey two ideas. First, because the word *authentic* implies "conforming to reality: TRUSTWORTHY" (Webster's *New Collegiate Dictionary*, 1987, p. 117), assessment of student performance should be trustworthy indicators of mathematical power (i.e., how well can students solve nonroutine problems?). Second, the term has been used for political purposes to imply that conventional tests are not trustworthy indicators of mathematical power. They are "inauthentic."

The seven chapters in this book have been prepared to raise a set of issues that scholars are addressing during this period of transition from traditional schooling practices toward the reform vision of school mathematics. The primary audience for this book consists of researchers in mathematics learning and teaching. It is anticipated that university professors and their graduate students will use the volume as a basis for discussion and potential studies during the next decade. However, because of the growing importance of assessment practices in schools, one would expect that many educational administrators, testing directors, and teachers of mathematics will find the chapters enlightening.

In chapter 1, Thomas Romberg and Linda Wilson raise a set of issues that they believe need to be addressed in order to build an assessment system for school mathematics. Susanne Lajoie presents in chapter 2 the argument about the need for authentic forms of assessment. In chapter 3, Edward Silver and Patricia Kenney focus on the importance of assessment information for making instructional decisions. Jan de Lange illustrates in chapter 4 a variety of authentic tasks used in The Netherlands to assess different levels of mathematical performance. In chapter 5, Robert Stake constructs a broad argument on the reasons why standardized testing is invalid in the current context of reform. Mark Wilson, in chapter 6, presents an alternative psychometric approach to mathematics assessment. Finally, in chapter 7, Elizabeth Graue reflects on the six previous chapters and points toward the need to extend the discussions and broaden our view of possibilities that need to be considered.

A number of people and organizations are responsible for making the publication of this book possible. The writing of individual chapters and the editing and preparation of the book were supported by the Office of Educational Research and Improvement of the U.S. Department of Education through the support of the National Center for Research in Mathematical Sciences Education. The Wisconsin Center for Education Research provided the ancillary services so necessary for this type of project. Andrew Porter, the director of the Wisconsin Center, and Jerry Grossman, business manager of the center, are thanked for their continued support. Joan Pedro is thanked for attending to many of the administrative details involved in its publication. Debra Torgerson is thanked for her work in typing and retyping manuscripts. Finally, special thanks go to Margaret Powell for the careful editing that contributed immeasurably to the clarity and quality of the writing of this book.

## REFERENCES

National Council of Teachers of Mathematics. (1989). *Curriculum and evaluation standards for school mathematics*. Reston, VA: Author.

Webster's. (1987). *Ninth New Collegiate Dictionary*. Springfield, MA: Merriam-Webster, Inc.

# 1 ❖ Issues Related to the Development of an Authentic Assessment System for School Mathematics

*Thomas A. Romberg and*
*Linda D. Wilson*

In 1990 the president of the United States and the National Governors Association announced their unprecedented agreement on national educational goals. For the nation to achieve those goals, it has become apparent that the American education system must be restructured. The strategy now being followed involves a series of steps to produce: a detailed set of content standards in English, mathematics, science, history, and geography; a set of standards describing how best to instruct students toward the attainment of each of those content standards; a set of procedures to assess student progress in meeting the content standards; and a set of standards to describe the responsibility of the professionals who will assist students in reaching those standards. The framework for restructuring schools via the specification of content standards, teaching standards, performance standards, and their interrelationships is shown in figure 1.1.

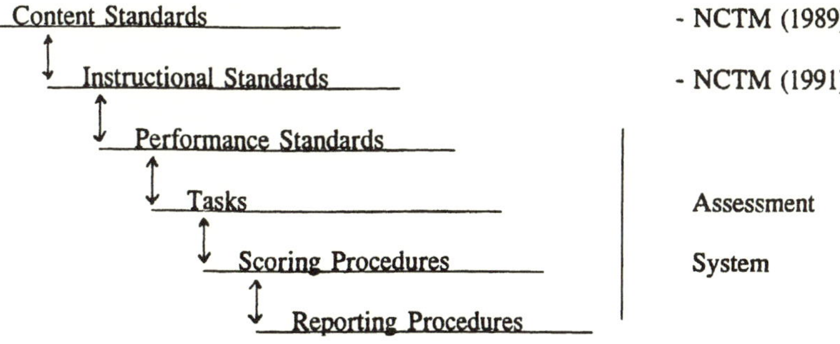



Figure 1.1. Relationships between content, instructional, and performance standards and an assessment system.

As shown in the figure, the initial stages of the framework have been addressed in mathematics. The *Curriculum and Evaluation Standards for School Mathematics* (National Council of Teachers of Mathematics, 1989) presents a consensual vision of the mathematical content that all students should have an opportunity to learn—the *content standards* in figure 1.1. Furthermore, the *Professional Standards for Teaching Mathematics* (National Council of Teachers of Mathematics, 1991) describes the means for assisting students to learn that content—the *instructional standards* in figure 1.1. In addition, some of the needed elements for the development of an assessment system are described in the *Curriculum and Evaluation Standards*. NCTM has now taken on the task of producing "assessment standards." They are due to be published in 1995.

The progress made in the content area of mathematics is being held up as exemplary for the other four disciplines. Nevertheless, a number of critical issues need to be addressed if the assessment system is to fit with the vision of the work done so far. In addition, a great deal of work remains to be done in the development of performance standards, assessment tasks, scoring, and reporting procedures. In this chapter, we identify a number of issues that we see as especially significant. Beginning with some assumptions upon which the reform movement in mathematics is constructed, we discuss what needs to be considered at each stage of development of an assessment system for mathematics for the result to be considered "authentic."

## ISSUE 1: UNDERLYING ASSUMPTIONS ABOUT THE NATURE OF MATHEMATICS

An authentic assessment system for school mathematics should begin with a vision about the nature of mathematics that is aligned with current thinking. An assumption underlying the development of any assessment system is that the responses a student makes to a set of test items or tasks will be a valid indicator of that student's understanding of some aspect of a domain of mathematics. There are three fundamental problems with this assumption. First, as Antoine Bodin (1993) has argued, one can never know what a student truly understands. One can only make inferences based on the responses a student makes to the tasks administered. This implies that the creation and selection of tasks is critical to the assessment process; in particular, they must reflect important aspects of mathematics a student has had an opportunity to learn. The second problem involves the reliability of the responses to those tasks so that a reasonable indicator of a student's understanding can be inferred. Together these lead to the final problem: What does one mean by *an understanding of mathematics?* The new and emerging answer to this question is at the heart of the calls to develop an "authentic" assessment system. To clarify this issue, we have chosen to describe the classical testing paradigm followed in the United States and point out its weaknesses with respect to current notions about the nature of mathematics.

Traditional norm-referenced standardized achievement tests for mathematics are created by following a particular measurement model. Such tests are made up of an assortment of independent, discrete questions that can be responded to quickly; all items are assumed to be equivalent; answers (usually derived by choosing among alternatives) are judged to be either correct or incorrect; and responses should be internally consistent, reflect important variations in responses between students, and be fair to all examinees. Such tests resolve the three problems mentioned previously by selecting or creating items that reflect specific concepts or procedures that appear in widely used textbooks; carefully considering a logical, hierarchical sequence of concepts and procedures; and having a group of teacher and mathematics educators judge their face validity. Reliability is established first by eliminating items that are too easy, too hard, or do not correlate with other items; and then an internal consistency coefficient is calculated. Finally, counting the number of correct responses on a test constructed in this manner is assumed to be a reasonable indicator of a student's knowledge, and differences in the number of correct responses among students is assumed to reflect differences in knowledge.

The calls to develop an "authentic" assessment system are based on the conviction that counting the number of correct answers to a series of brief questions contradicts current views of mathematics as an intellectual discipline. Ernest (1991), for example, argues that mathematics cannot be described by a single unique hierarchical structure and that mathematics cannot be represented as a set of discrete knowledge components. The mathematician William Thurston (1990) uses the metaphor of a tree to describe mathematics: "Mathematics isn't a palm tree, with a single long straight trunk covered with scratchy formulas. It's a banyan tree, with many interconnected trunks and branches— a banyan tree that has grown to the size of a forest, inviting us to climb and explore" (p. 7). A valid system for assessment in mathematics must reflect these notions—that mathematics is a set of rich, interconnected ideas. To be in line with current thinking, it must view mathematics as a dynamic, continually expanding field of human creation, a cultural product (Ernest, 1988).

In the NCTM *Standards* (1989), the development of mathematical power is presented as the central goal of school mathematics. *Mathematical power* is defined as the ability to "explore, conjecture, and reason logically, as well as the ability to use a variety of mathematical methods effectively to solve nonroutine problems" (NCTM, 1989, p. 5). The term is based on a recognition that mathematics is more than a static collection of discrete concepts and skills to be mastered. Doing mathematics includes such dynamic and integrative activities as discovering, exploring, conjecturing, sense making, and proving. Students who possess mathematical power should be able to investigate and reason, communicate ideas, and take real contexts of problems into account. The descriptive verbs used in the *Standards* evoke images of mathematics as a progressive human activity.

If one considers mathematics to be a static, linearly ordered set of discrete facts, then the logical choice for a valid assessment system is the traditional standardized achievement test. On the other hand, if one views mathematics as a dynamic set of interconnected, humanly constructed ideas, then the assessment system must allow students to engage in rich activities that include problem solving, reasoning, communications, and making connections.

## ISSUE 2: UNDERLYING ASSUMPTIONS ABOUT THE LEARNING OF MATHEMATICS

It is critical that in addition to being based on certain beliefs about the nature of mathematics an assessment system be built on current views of learning mathematics. A recent study by Shepard (1991) showed that approximately half of all district testing directors in the United States hold beliefs about the alignment of tests with curriculum and teaching that are based on behaviorist learning theory, which requires sequential mastery of constituent skills and behaviorally explicit testing of each learning step. Such a learning theory was prevalent for several decades, but is now out of date with current research.

Indeed, as Romberg, Zarinnia, and Collis (1990) noted, the values and forces that dominated mathematics education for the past century (e.g., behaviorism) are embedded in the theoretical structures of prevailing methods of assessment. Tests built on behavioral objectives and a content-by-process matrix are based on behaviorist ideas about learning: that content can be broken down into small segments to be mastered by the learner in a linear, sequential fashion.

Yet a substantial body of evidence from cognitive psychology shows this hierarchical model of learning to be obsolete. The metaphor of the learner as a passive absorber of linearly ordered bits of information is contradicted by research findings from psychology. Resnick (1987) has argued that learning does not occur by passive absorption alone, but rather in many situations learners approach a new task with prior knowledge, assimilate new information, and construct their own meanings. Ernest (1991) has shown that the uniqueness of learning hierarchies in mathematics is not confirmed theoretically nor empirically. Furthermore, he argues against the notion that concepts in mathematics can be either "possessed" or "lacking" in a learner.

The shift in learning theory can best be summarized as a move from behaviorism to constructivism. Though there is not total agreement in the mathematics education community about exactly what a "constructivist" theory of learning entails, Peterson (in press) has described four basic assumptions that form the foundation for current theory, research, policy, and practice in mathematics education:

- Learners are knowledgeable "sense makers."
- Learning involves the negotiation of shared meaning.
- Knowing is contextualized or situated.
- Assumptions about knowledge influence learning.

A more appropriate metaphor for learning may be an image that is gradually brought into sharper focus as the learner makes connections, or perhaps like a mosaic, with specific bits of knowledge situated within some larger design that is continually being reorganized or redesigned in an organic manner. In either case, the emphasis is on knowing, rather than "knowing that." The *Standards* express it as a process: " 'Knowing' mathematics is 'doing' mathematics. A person gathers, discovers, or creates knowledge in the course of some activity having a purpose. This active purpose is different from mastering concepts and procedures. We do not assert that informational knowledge has no value, only that its value lies in the extent to which it is useful in the course of some purposeful activity. It is clear that the fundamental concepts and procedures from some branches of mathematics should be known by all students. Established concepts and procedures can be relied on as fixed variables in a setting in which other variables may be unknown. But instruction should persistently emphasize 'doing' rather than 'knowing that'" (NCTM, 1989, p. 7). Assessment, then, should be based on a view of the learning of mathematics as a socially constructed process, not a fixed hierarchy of skills and concepts to be mastered.

## ISSUE 3: THE NEED FOR NEW PSYCHOMETRIC MODELS

"It is only a slight exaggeration to describe the test theory that dominates educational measurement today as the application of twentieth century statistics to nineteenth century psychology" (Mislevy, 1990, abstract). When Mislevy wrote those words in 1990 he was calling for the field of psychometrics to "catch up" with the advances in cognitive psychology. As noted earlier in Shepard's (1991) work, many psychometricians are still operating under theories of learning and measurement that are out of date. New knowledge of how learning takes place must be accounted for in psychometric theory. "Learners become more competent not simply by learning more facts and skills, but by reconfiguring their knowledge; by 'chunking' information to reduce memory loads; and by developing strategies and models that help them discern when and how facts and skills are important. Neither classical test theory nor item response theory (IRT) is designed to inform educational decisions conceived from this perspective" (Mislevy, Yamamoto, & Anacker, 1992). Just as an assessment system must be built on current learning theory, so must the psychometric measurement theories that support such a system be designed with cognitive psychology as its base. Fortunately, work toward new theories of testing is being accomplished, as some psychometricians are now realizing. Wilson (1992) expressed the need this way: "The consequence of this view of learning [constructivism] is that we can no longer use an atomistic model for assessment. We must assess the level of complexity of student understanding, not just the number of facts that students can pick out of a multiple-choice test" (p. 123). New models, such as those described by Mark Wilson in chapter 7, are being constructed that begin to capture more of the complexity of learning than was allowed for by standard test theory. Although no new model claims to describe all of the nuances of current learning theory, with the support of more powerful technologies progress is being made (Glaser, Lesgold, & Lajoie, 1987; Mislevy, Yamamoto, & Anacker, 1992). It is critical that an authentic assessment system take such work into account in its design.

## ISSUE 4: ALIGNMENT WITH THE REFORM CURRICULUM

As described in figure 1.1, the first stages of the building of an assessment system for mathematics—that is, setting content and instructional standards—has been accomplished. Consensus has been reached in the mathematics education community about the content that all students should be given the opportunity to learn and the pertinent means of instruction. As the next four stages (setting performance standards, developing tasks, adopting scoring, and reporting procedures) are undertaken, it is critical that the outcomes be in alignment with the conceptualizations of curriculum and instruction set forth in the *Standards*.

The *Standards* are built on a set of assumptions about the nature of mathematics, about learning, and about teaching. As described earlier, mathematics is viewed as a progressive human activity. To know mathematics is to engage in the activities of doing mathematics, such as conjecturing, sense making, and communicating mathematical arguments. Another fundamental assumption is that school mathematics is not solely for the elite, but for *all* students. All students come to school with certain mathematical concepts already forming, and the role of the teacher is to build on that knowledge so that students gain increasing mathematical power. The teacher's role is no longer that of a deliverer of knowledge, but that of a guide and facilitator for student growth.

To be in accord with the work completed thus far on the mathematics curriculum and methods of instruction, the next stages in the development of an assessment system must take these fundamental assumptions into account. This implies, for example, that performance standards should be based on students solving nonroutine problems rather than performing conventional computational procedures. Assessment tasks should allow students the opportunity to demonstrate their mathematical power. They should require the active engagement of students in doing mathematics rather than making a passive response to routine questions. Also scoring and reporting methods should be designed to inform individual students about their own learning rather than to rank students in groups.

### ISSUE 5: SPECIFICATION OF PERFORMANCE STANDARDS

The NCTM *Curriculum and Evaluation Standards* (1989) describe what students should have an opportunity to learn. But to establish an assessment system for school mathematics that is aligned with that vision, performance standards must be set that will describe what students are supposed to be know and be able to do in mathematics. Making the connection between curriculum standards and performance standards is a difficult task, but one that needs to be confronted.

A conventional approach to testing specifies both content and levels of performance; it then crosses them to form a "content-by-process" matrix. Although this approach affords test developers the assurance that each content area of mathematics and each level of performance (or type of process required) will be "covered" by the test, the design may also work against the assumptions about mathematics mentioned earlier. That is, separating mathematics into individual cells of "Number" as content and "Computation" as process, for example, sets up a situation for test writers to write items that fit neatly into those cells. The design does not easily allow for items that require more than one content area or more than one process in their solution. As an example, consider this item, similar to items found on recent standardized achievement tests (Romberg & Wilson, 1992):

> Which is best to use to find an estimate for 791 ÷ 19?
>
> A. 700 ÷ 10
> B. 700 ÷ 20
> C. 800 ÷ 10
> D. 800 ÷ 20

This item would fit (all too neatly) into a content area of Number and a process area of Computation/Estimation. When tests are composed primarily of items like these, the underlying assumption is that mathematics is a collection of discrete content areas and that the doing of mathematics occurs in a separate, compartmentalized, hierarchical fashion.

For a task to be considered "authentic," it should not easily fit into neat categories of single content areas and single processes. Solving nonroutine problems usually involves multiple processes and cuts across mathematical domains. Making connections necessarily involves blurring the lines between content and processes. The task (NCTM, 1989, p. 141) presented in figure 1.2 illustrates the interconnectedness of problem solving, communication, and reasoning and involves the content areas of geometry and discrete mathematics.

To build an assessment program in alignment with the NCTM *Standards* implies that all the elements of the program incorporate the four major strands of emphasis: problem solving, communication, reasoning, and connections. This does not imply that every task would necessarily include all four standards, but that the assessment program as a whole would incorporate all four of them at each level. It would also be essential that these four strands not be separated into distinct "process categories," but rather that the items reflect their necessary overlap and interrelatedness.

## ISSUE 6: DEVELOPING AUTHENTIC TASKS

Increasing attention is being given to notions of "authentic assessment." Definitions or criteria for authentic assessment are being developed that are built on the framework of the reform curriculum in mathematics education. For an assessment system to be considered "authentic," it must acknowledge these criteria.

Archbald and Newmann (1988) consider three criteria to be critical to authentic assessment tasks: (1) disciplined inquiry; (2) integration of knowledge; and (3) value beyond evaluation.

Nine robots are to perform various tasks at fixed positions along an assembly line. Each must obtain parts from a single supply bin to be located at some point along the line. Investigate where the bin should be located so that the total distance traveled by all of the robots is minimal.

Figure 1.2. An example of a task for grades 9–12 (Reproduced with permission from *Curriculum and Evaluation Standards for School Mathematics*, copyright 1989, by National Council of Teachers of Mathematics, p. 141).

*Disciplined inquiry* refers to the production of new knowledge, such as that created by scientists or historians. It depends on prior conceptual and procedural knowledge, it develops in-depth understanding of a problem, and it "moves beyond knowledge that has been produced by others" (p. 2). *Integration of knowledge* means that authentic tasks must consider the content as a whole, rather than as a collection of knowledge fragments. Students must "be challenged to understand integrated forms of knowledge" and "be involved in the production, not simply the reproduction, of new knowledge, because this requires knowledge integration" (p. 3). The third criterion, *value beyond evaluation*, refers to the idea that authentic tasks should possess attributes that make them worthwhile activities beyond their use as evaluative tasks. An example would be a task that results in discourse, an object, or a performance. An authentic task might also have value for the collaborative opportunities it provides.

Although these criteria are more broadly based, Lajoie in chapter 2 develops a set of criteria for authentic assessment specifically in mathematics. This framework does not contradict the more general notions of Archbald and Newmann, but makes more explicit the ways in which the content of mathematics influences the design of assessment tasks. It is built on two primary foundations: the NCTM *Standards* (1989) and current learning theory. The *Standards* are predicated on two basic assumptions: the first, that knowing mathematics is doing mathematics, and the second, that there should be four goals for school mathematics content—problem solving, communication, reasoning, and connections. From current learning theory, Lajoie (1991) chooses situated cognition and social constructivism to form a foundation for the definition of authentic assessment.

Building on concepts of mathematics in the *Standards* and on learning theory, Lajoie defines seven principles for a definition of authentic assessment:

1. It must provide us with multiple indicators of the learning of the individual in the cognitive and conative dimensions that affect learning. The cognitive dimensions include content knowledge, how that knowledge is structured, and how information is processed with that knowledge. The conative dimensions should address students' interest in and persistence on tasks, as well as their beliefs about their ability to perform.

2. It must be relevant, meaningful, and realistic. It must be instructionally relevant, as indicated by its alignment with the

NCTM *Standards*. It must relate to pure and applied tasks that are meaningful to students and that provide them with opportunities to reflect, organize, model, represent, and argue within and across mathematical domains.

3. It must be accompanied by scoring and scaling procedures that are constructed in ways appropriate to the assessment tasks.

4. It must be evaluated in terms of whether it improves instruction, is aligned with the NCTM *Standards*, and provides information on what the student knows.

5. It must consider racial or ethnic and cultural biases, gender issues, and aptitude biases.

6. It must be an integral part of the classroom.

7. It must consider ways to differentiate between individual and group measures of growth and to provide for ways of assessing individual growth within a group activity (pp. 30–31).

This set of criteria, in defining authentic assessment in mathematics, could serve as a guideline for an authentic assessment system for school mathematics. It incorporates current learning theories and emphasizes the necessary alignment with the reform curriculum.

The format of authentic tasks may vary. In fact, in keeping with the need for multiple sources of information, no assessment system should be limited to a single form. In chapter 4, de Lange discusses the various formats of mathematical tasks, from multiple choice to portfolios, and sheds some light on what is meant by an "open" item.

## ISSUE 7: MEASURING STATUS, GROWTH, OR A COMBINATION

It is clear that all forms of assessment, including traditional standardized tests, are designed to measure the present status of student thinking. Traditional measuring instruments were created to yield highly reliable scores on a single dimension, with the ultimate purpose of linearly ranking students on that dimension. This factorylike image of education belongs to an earlier age when behaviorist theories of learning held sway. Constructivist approaches to assessment require greater emphasis on a developing picture of individual growth. The emphasis has shifted from an industrial model of quality control to an effort to describe an individual's attainment of mathematical power. There is a need for more than status information; instead of a static score, what is needed are profiles of growth over time.

A single score, although useful for ranking students at a fixed point in time, places the emphasis of the assessment on the measure used. When assessment is seen an a means to understanding a student's growth over time, the emphasis shifts to the process of learning. Such a change vastly increases the utility of the information gained for all the audiences involved. Students, no longer limited to information focusing only on comparisons with peers on a single measure, gain an understanding of their own learning. Measures of growth over time are immensely consequential for teachers in planning instruction. As students gain more sophistication in their problem-solving strategies, assessment can best inform students and teachers by describing growth in that ability longitudinally. Parents and administrators also gain a deeper understanding of student learning when information is provided that goes beyond a single static score.

### ISSUE 8: SCORING—BY WHOM AND IN WHAT FORM?

Scoring on traditional standardized tests has historically been done by machine, which works well with multiple-choice, single-answer items. But different forms of assessment with open-response items, for example, require professional judgment to score. The issue then is, Can we trust teachers to reliably mark their own students' work, and can they be trained to do so? Other countries have struggled with this issue as well and have responded with a variety of strategies and results.

In The Netherlands, an experiment was conducted to test the reliability of teachers' judgments. Fifteen teachers were asked to score the work of five students on an extended open-ended task. The teachers were given no information on the students, no information on the results of each student's previous work, and no indication on how to score the tasks, other than to use a ten-point scale. The responses yielded high interrater agreement among the fifteen teachers. In 81 percent of the cases, two scores of a student's task lay within 1.0 points of each other. When all the averages of any two scores and the average of all scores (considered the "correct grade") were calculated, roughly 90 percent of the averages of any two scores lay within a half point of the "correct" grade (de Lange, 1987, pp. 209–220). Presumably, teachers given a rubric (or entrusted to develop one) and trained in its use would yield even higher rates of agreement.

There are models in other countries of statewide mandated examinations that rely on the expertise of classroom teachers for scoring, while incorporating strategies for external verification. In Victoria, Australia, an external verification process has been used for the Certificate of Education exams, which are composed of four different kinds of tests, ranging from multiple choice to extended projects. The process, which checks the scoring of teachers and ensures reliability, has been found to have significant professional development benefits for the teachers involved as well. Before the examinations are undertaken, there is a training activity at which student work from previous years is examined in an effort to bring teachers to a common understanding of the desirable attributes of student reports and the criteria for assigning grades. After teachers have assigned grades, they submit typical examples of student work and difficult or ungraded cases to a regional review panel. Eventually, examples of work from each region are forwarded to a statewide panel for review. The review panels suggest such alterations to grades as seem appropriate, and teachers may then reassess students' work, taking the panel's advice into account.

On a final verification day, all students' work is brought to a regional meeting. Teachers are divided into verification teams under the direction of a review panel member. These teams reassess a number of reports selected by the panel member. Teachers do not reassess work from their own schools. If significant variation is found between the initial and second grades, further sampling is done and a whole class might be re-marked by at least two members of the verification team. In the trials, the grades assigned by verification teams in this way have been "remarkably consistent because sufficient professional development had taken place to ensure a common understanding of the grading process. The grades resulting from this process are as comparable as one can reasonably expect short of double marking the entire collection of reports" (Stephens & Money, 1991, p. 5).

Wilson (1992) has offered another strategy for combining the expertise that teachers have regarding individual students in their class with the more tightly controlled ratings of external examiners. He offers an analytical scheme that combines the two types of scores statistically in a manner that can give credence to both types of assessment.

The Victoria example illustrates that such processes, which rely on teachers to score the tests with sufficient checks in place, can be put into practice and that there are supplementary benefits in the professional development of teachers. Teachers who meet to discuss the rubrics and expectations for student work can develop a common language for the assessment of these tasks, but they can also take back to their classrooms a clearer vision of the kinds of mathematical activities valued. Teachers who cooperate to create and use agreed rubrics for evaluating student work gain valuable experience at using alternative forms of assessment. In a context in which they are reasonably sure of the reliability of their judgments against those of other teachers, they can in turn feel more confident in their own grading for instructional decision making. At the same time, formal rubrics and strategies for accomplishing interjudge agreement can support the development of alternative assessment tasks, prompt efforts to improve the mathematics curriculum to enable students to achieve those goals, and help instill public confidence in the use of school-based information for accountability.

## ISSUE 9: MAKING REPORTS OF RESULTS UNDERSTANDABLE TO THE PUBLIC

Results of student performance on an examination need to be reported to several audiences: students, parents, teachers, administrators, and policy makers. The form and substance of these reports will necessarily vary according to the audience. Nevertheless, it is essential that they are designed to be easily understood by their constituents, at the same time that they preserve the richness of the information. It would do little good to replace traditional tests with nontraditional formats if the conceptually abundant information gathered is collapsed into a single numerical score or left in an uninterpreted form.

The first question that must be addressed in the design of reports is, What kinds of information does each audience (students, parents, teachers, administrators, or policy makers) need to make informed decisions? Because even the unit of analysis is different for each audience, the information required will likewise vary. Once that element is decided, the appropriate data sources and means of analysis can be determined.

One model for reporting is being developed by Lesh, Lamon, Gong, and Post (1992). Their "learning progress maps" are computer based, interactive, multidimensional, and decision specific, yet relatively simple in design. The maps report student progress along three dimensions. The vertical axis represents the most important conceptual models and reasoning patterns that students are encouraged to construct at a given grade level. On the horizontal axis are the basic mathematical strands (such as patterns, quantities); and the depth axis corresponds to the increasing structural complexity of the underlying conceptual systems. The result is a visual image of student learning, in the form of peaks and valleys.

Another proposed framework for reporting student progress in mathematics learning (Romberg, 1987) utilizes Vergnaud's notions about "conceptual fields." The idea is that, rather than breaking down mathematics into two dimensions of content and processes, a vast number of different forms of problem situations in mathematics can be represented by a small number of symbols and symbolic statements. For example, the related mathematical concepts of addition and subtraction of whole numbers has been defined by Vergnaud as the conceptual field "additive structure." Developing such fields can yield a map of a domain of knowledge. Such maps could free test constructors from the bind of filling in cells of a matrix.

A common challenge to current efforts at assessment reform is the development of profiles of student learning that are meaningful, concise, valid, and reliable, at the same time that they are based on a framework built on current notions of learning mathematics. One such attempt is the scheme offered by Zarinnia and Romberg (1990), shown in figure 1.3. The subcategories under *Doing Mathematics* are not the usual mathematical terms, such as space or logic. The idea is to try to capture a more realistic picture of genuine mathematical activity. The language of the chosen terms emphasizes mathematics in terms of active engagement, creative reflection, and productive effort.

| Doing Mathematics | Representing & Communicating Mathematics | Mathematical Community | Mathematical Disposition |
|---|---|---|---|
| Locating | Mental and | Individual | Valuing math |
| Counting | Represented | activity | confidence |
| Measuring | Facility in | Collaborative | Beliefs about the |
| Designing | Communicating: | activity | mathematical |
| Playing | verbally | | enterprise |
| Explaining | visually | | Willingness to |
| | graphically | | engage and |
| | symbolically | | persist |

Figure 1.3. Recommended reporting categories (Zarinnia & Romberg, 1990, p. 31).

What each of the three alternatives described briefly here has in common is an attempt to respond to current thinking about the learning of mathematics. No longer will a single numerical score suffice to describe the complex processes involved in engaging in the kinds of mathematical activity described by the *Standards*. A reporting system that seeks to support, not undermine, an authentic assessment system will have to be sophisticated enough to embrace a more complex view of the learner and an enlightened view of what it means to do mathematics at the same time that it generates information that is useful to students, teachers, parents, administrators, and the public for decision-making purposes.

The list of issues discussed in this chapter is not meant to be exhaustive, nor have we tried to resolve all of the dilemmas of each one. The other chapters of this book will elucidate some of them. Our intent is to bring to light some of the important matters that need to be addressed at each of the stages in building an assessment system for mathematics, from the initial assumptions made about mathematics and learning to the reporting schemes used. Only with careful attention to each stage of the process, and a clear vision of the overall framework, can a coherent authentic assessment system be constructed.

## REFERENCES


Archbald, D., & Newmann, F. (1988). *Beyond standardized testing: Assessing authentic academic achievement in the secondary school*. Reston, VA: National Association of Secondary School Principals.

Bodin, A. (1993). "What does to assess mean? The case of assessing mathematical knowledge." In M. Niss (ed.), *Investigations into assessment in mathematics education: An ICMI study*, pp. 113–141. Dordrecht: Kluwer Academic Publishers.

de Lange, J. (1987). *Mathematics, insight, and meaning*. Utrecht: OW & OC.

Ernest, P. (1988). "The impact of beliefs on the teaching of mathematics." Paper presented at the International Congress on Mathematics Education VI, Budapest.

————. (1991). *The philosophy of mathematics education*. London: Falmer Press.

Glaser, R., Lesgold, A., & Lajoie, S. (1987). "Toward a cognitive theory for the measurement of achievement." In R. Ronning, J. Glover, J.C.

Conoley, & J. Witt (eds.), *The influence of cognitive psychology on testing and measurement: The Buros-Nebraska Symposium on Measurement and Testing*, vol. 3. Hillsdale, NJ: Lawrence Erlbaum Associates.

Lajoie, S. (1991, October). "A framework for authentic assessment in mathematics." *NCRMSE Research Review*: 6–11.

Lesh, R., Lamon, S., Gong, B., & Post, T. (1992). "Using learning progress maps to improve instructional decision making." In R. Lesh & S. Lamon (eds.), *Assessments of authentic performance in school mathematics*, pp. 343–378. Albany: SUNY Press.

Mislevy, R. (1990). "Foundations of a new test theory." In N. Frederiksen, R. Mislevy, & I. Bejar (eds.), *Test theory for a new generation of tests*, pp. 19–40. Hillsdale, NJ: Lawrence Erlbaum Associates.

———, Yamamoto, K., & Anacker, S. (1992). "Toward a test theory for assessing student understanding." In R. Lesh & S. Lamon (eds.), *Assessments of authentic performance in school mathematics*, pp. 293–318. Washington, DC: American Association for the Advancement of Science.

National Council of Teachers of Mathematics (NCTM). (1989). *Curriculum and evaluation standards for school mathematics*. Reston, VA: Author.

———. (1991). *Professional standards for teaching mathematics*. Reston, VA: Author.

Peterson, P. (In press). "Learning and teaching mathematical sciences: Implications for inservice programs." In *Teacher enhancement for elementary and secondary science and mathematics: Status, issues, and problems*. Arlington, VA: National Science Foundation.

Resnick, L. (1987). *Education and learning to think*. Washington, DC: National Academy Press.

Romberg, T. A. (1987, November). "The domain knowledge strategy for mathematical assessment" (Project Paper #1). Madison, WI: National Center for Research in Mathematical Sciences Education.

——— & Wilson, L. (1992). "Alignment of tests with the *Standards*." *Arithmetic Teacher* 1, no. 40, 18–22.

———, Zarinnia, A., & Collis, K. (1990). "A new world view of assessment in mathematics." In G. Kulm (ed.), *Assessing higher order thinking in mathematics*, pp. 21–38. Washington, DC: American Association for the Advancement of Science.

Shepard, L. (1991). "Psychometricians' beliefs about learning." *Educational Researcher*, 20, no. 7:2–16.

Stephens, M., & Money, R. (1991, April). "The range of performance assessed: The interaction between new developments in assessment for the senior secondary years and curriculum change in mathematics." Paper presented at the International Commission on Mathematics Instruction, Calonge, Spain.

Thurston, W. (1990). "Letters from the editors." *Quantum*, 1, no. 1:7–8.

Wilson, M. (1992, January). "Measurement models for new forms of student assessment in mathematics education." Paper presented at the meeting of the Australian Council for Educational Research, Melbourne (Models of Authentic Performance Working Group paper). Madison, WI: National Center for Research in Mathematical Sciences Education.

Zarinnia, A., & Romberg, T. A. (1990, February). *A framework for the California assessment program to report students' achievement in mathematics*. Madison, WI: National Center for Research in Mathematical Sciences Education.

## 2 ❖ A Framework for Authentic Assessment in Mathematics

*Susanne P. Lajoie*

Vast differences exist between the tasks learned in school mathematics and those tasks mathematicians or users of mathematics actually carry out (Lampert, 1990; Pollak, 1987; Resnick, 1988). How we learn inside the classroom is different from how we learn outside of the classroom (Resnick, 1987). Resnick elaborates that the focus inside the typical American classroom is on what the individual learner can accomplish independent of the group or tools for learning such as calculators. In contrast, outside-of-classroom learning situations often are group situations, where knowledge must be shared and where tools are often available to enhance or extend our knowledge. Inside the classroom students are taught to manipulate symbols and abstract principles, but outside the classroom learning often is concrete and situated in the context in which it will be used. Given the differences between the two, the term *authentic* has been used to suggest that some classroom activities are lacking in realism and to conjure up an image of an alternative approach.

Requests for more authentic classroom activities have led to requests for authentic forms of assessment. These requests have come from several populations—ranging from students, teachers, and district and state personnel to a national agenda on the integration of instruction and assessment. Although the rhetoric is convincing, the images of authentic activities and assessment are still imprecise. This chapter is written to stimulate discussion on ways that *authentic assessment* can be operationally defined in the area of school mathematics.

The distinctions between in-school and out-of-school learning have implications for defining authentic forms of assessment. In considering these distinctions, we must also consider whether a framework for authentic assessment should incorporate a guide for authentic instructional activities in the classroom. Should we be concerned with mathematical knowledge that transfers to everyday uses of mathematics or should we consider authentic mathematics as something mathematicians do in their domain? Several researchers have demonstrated that incorporating everyday uses of mathematics into instruction improves both interest and performance (Fong, Krantz, & Nisbett, 1986; Mosteller, 1988). One research strategy has been to identify the learners' informal knowledge of mathematics (statistics in particular), which comes from the learners' everyday experiences, and formalize this knowledge with appropriate instruction (Fong et al., 1986). It is still an empirical question as to what mathematicians do in their domains and how such knowledge could be modeled in the traditional classroom.

Part of what makes mathematics authentic is the fact that mathematics is essential to the needs of our rapidly changing society. Computer technology has become more the norm in the workforce and thus there is a greater need for mathematically literate workers. Mathematical literacy can be measured as the ability to understand the complexities of technologies, to be able to communicate and ask questions, to assimilate unfamiliar information, and to work cooperatively in teams. These skills are skills for lifelong learning. The fact that our society demands job mobility ensures the need for flexibility and adaptability, where individuals must be capable of learning new information quickly and commuicating what they understand or do not understand. Adaptation can be fostered by providing multiple learning contexts that encourage students to value mathematical interpretations in a variety of interrelated experiences. Communication can be fostered if schooling helps students learn the language of mathematics and if schooling provides opportunities to conjecture and reason (Lampert, 1990; Resnick, 1987). The Department of Labor and the business community are recommending that we reconsider the type of criteria that certify high school graduates and perhaps include certificates that reflect the skills that are required in the workforce (see Whetzel, 1992, for a review of the Secretary's Commission on Achieving Necessary Skills [SCANS] test).

My primary focus in defining authentic assessment in mathematics is to provide a robust perspective of the individual learner's understanding of mathematics. Several audiences are considered as I define worthwhile mathematical tasks from the mathematics educator's perspective, followed by a description of two interrelated theoretical perspectives on authentic activities as described in the literature on situated cognition and social constructivism.

## WORTHWHILE MATHEMATICAL TASKS

The National Council of Teachers of Mathematics' *Standards* (1989) present goals for worthwhile or essential mathematics that are designed to make students at each level of schooling mathematically powerful. The first four of the standards represent overarching goals that should be considered for all mathematics content at all levels. Any specific mathematical content, according to these four, should be designed to provide students with opportunities for mathematical problem solving, communication, reasoning, and making connections. These goals must be translated into tasks that exemplify authenticity. Only then can a framework for authentic assessment be developed.

A definition of an authentic mathematical activity emerges from the general assumptions of the NCTM *Standards*. One assumption is that knowing mathematics is doing mathematics. *Doing mathematics* refers to gathering and discovering knowledge in the course of solving genuine problems where knowledge emerges from experiences that are challenging but solvable. One way to increase such opportunities is to provide students with experiences in building mathematical models, structures, and simulations across multiple disciplines. Model building and discovering mathematical patterns is often a dynamic constructive process. Technology can be used to facilitate these cognitive processes as well as record them. It can be used to assess developmental changes in reasoning, hypothesis formulations, verifications, and revisions. Technology can also serve as a medium for instructional manipulation where small changes in the instructional environment may account for changes in the learners' acquisition of knowledge.

## PROBLEM SOLVING

Activities that give students experience with problem solving can emerge from problem situations. These situations can be used to motivate students and serve as a context in which information is learned and knowledge is re-created across grades. For example, the use of statistics in everyday situations is apparent in any newspaper. Simply turn to the weather forecasts for predictions, the sports section for rankings, and the business section for percentages. Pereira-Mendoza & Swift (1981) discuss ways in which the real-world context can be brought to the mathematics classroom by using newspaper articles that feature statistics as a starting point for discussions to provoke statistical reasoning. Another way to make statistics more authentic is to use meaningful data sets as opposed to "cooked up" ones (Singer & Willett, 1990; Tanner, 1985). By using real data, students come to understand the phenomenon and see how and why statistics are useful in authentic contexts. Although "real" data are often complex, messy, and frequently culturally based, their use does provide opportunities for multiple strategies and solutions to evolve (Zarinnia & Romberg, 1992). Real-world problems can provide more freedom for learners to pursue questions that reflect their personal interests.

Problem solving with mathematics involves modeling the problem and formulating and verifying hypotheses by collecting and interpreting data using pattern analysis, graphing, or computers and calculators. Technology is a powerful tool; it permits learners to manipulate data and see the consequences of their work in a few seconds. Some progress is being made in examining the effectiveness of graphing calculators and the use of computers in facilitating mathematics performance. Wainer's (1992) research on graphical displays provides us with insight into mathematical problem solving. One insight is that a graph, if properly drawn, can facilitate the discovery of relationships in the data and often answer both simple and complex questions. Wainer describes how graphs can be used to assess different levels of reasoning when paired with the appropriate questions. Bertin (1973) developed such a set of questions. For example, an elementary question may involve finding one data point, whereas an intermediate question could involve finding trends among multiple points in the data. The most sophisticated question would involve testing the learner's understanding of the deep structure of the data, which might include identifying multiple trends or understanding the overall picture. Intuitively, assessment of learning through questioning seems possible. However, Wainer cautions that questions and graphs alone may not result in appropriate assessments of the learner's reasoning capability, simply because the graph may be poorly constructed and thus misrepresent what the learner can or cannot see in the data.

The use of computers to promote mathematical problem solving is becoming more popular. Reed (1985) examined the effects of computer graphics on improving estimates in algebra word problems. He varied the instructional environment so that students either learned by viewing computer simulations or learned by doing, where they saw the consequences of their estimates on the computer screen (visual feedback, using computer graphics).

He found that simply viewing graphics did not improve performance. Viewing by doing, or utilizing visual feedback, was more successful at improving performance. However, the results were somewhat mixed and he concluded that certain displays were effective for certain tasks and not for others and that learning by coaching would be more effective. In other words, a learning-by-doing condition with intelligent feedback provided by the computer in the context of the student-generated estimates might have a stronger influence on performance.

Problem-solving activities need to include those that apply mathematics to the real world and those that arise from the investigation of mathematical ideas. Traditional curricula have emphasized mathematical ideas. The impetus for developing real and relevant problems stems from the need to contextualize mathematical concepts in a concrete rather than abstract manner. Real-world situations facilitate connectiveness of knowledge, understanding of contexts and goals, and fewer distractions (see chapter 4, this book). Equity issues must be considered in the development of these real-world problems so that cultural biases are not introduced. In addition to including applied and pure mathematical problem types, problem representations should be varied to provide for individual differences—that is, verbal, numerical, graphical, geometrical, or symbolic—and to permit several ways of reaching a solution.

### COMMUNICATING

Communicating about mathematical ideas permits students to synthesize information about the ideas. There are a variety of modes of communication, including reading, writing, discussion, and listening, as well as concrete, pictorial, graphical, and algebraic methods. Activities that require students to communicate about mathematics provide them with opportunities to reflect on and clarify their own thinking and to develop a communal understanding of mathematical ideas and notations.

Students need opportunities to present ideas using language to ensure that they understand words and their definitions and meanings. Teachers who structure classes to encourage communication provide students with opportunities to validate their thinking about mathematics. They can foster communication by asking questions, posing problems, or asking students to develop problems. Different levels of communication can be obtained by interviewing individual students, by using small groups, or by classroom discussions. These levels permit students to ask questions, discuss ideas, offer constructive criticism, and summarize discoveries in writing. Cultural and gender differences should be considered by those structuring activities to encourage communication.

## REASONING

Mathematics involves both inductive and deductive reasoning. Inductive reasoning is associated with mathematical creativity or invention. Deductive reasoning involves understanding the premises of a mathematical problem and thinking logically using the information given. Challenging problem situations can provide opportunities for students to develop mathematical reasoning in a variety of contexts. The maturation of mathematical reasoning is a long process. Special developmental differences in reasoning, especially in grades 5–8, where students make the transition from concrete to abstract reasoning, must be planned for. The development of mathematical reasoning could be facilitated in both instructional and assessment settings if appropriate prompts were made available to the learners: Why is this true? What if you changed this? Do you see a pattern?

## MAKING CONNECTIONS

A curriculum that integrates a broad range of mathematical topics rather than treating each topic in isolation is a connected curriculum. Number concepts, computation, estimation, functions, algebra, statistics, probability, geometry, and measurement become more useful to students when treated in an integrated fashion. Students can be helped to make connections between the topics if they are provided with contexts that require their integration when solving problems. Bryant (1984) discusses the interconnections between geometry, statistics, and probability. He describes how each topic has its own language for expressing meaning by repeating examples in different words and emphasizing the equivalence of the various means of expression that will provide students with assistance in drawing the necessary interconnections. It is not enough, however, to provide connections among mathematical topics; the connection of mathematics with other topics and with such disciplines as science, music, and business is also necessary

(Bransford et al., 1988; Rosenheck, 1991). Teachers from other disciplines can help to identify the mathematical ideas that can be explored in their domains. Geography, for example, provides opportunities for the use of scaling, proportions, ratio, similarity, and other mathematical ideas. Genetics, as a scientific discipline, provides ample opportunities for the application of mathematics, especially statistics and probability (Ballew, 1981). Ballew describes how basic statistical techniques of gathering and organizing data can be used to explain heredity. Using mathematics in specific contexts promotes attitudes of inquiry and investigation as well as sensitivity to the interrelationships between formal mathematics and the real world.

Problem solving, communicating, reasoning, and making connections can be seen as curriculum goals that permeate the entire mathematics curriculum. Specific content areas also need to be addressed: number and number relations, number systems and number theory, computations and estimation, patterns and functions, algebra, statistics, probability, geometry, and measurement. In reviewing what the NCTM *Standards* (1989) deem worthwhile mathematical activities, it is important to realize that a single assessment of such activities will not provide a complete picture of a student's intellectual growth. Furthermore, different types of assessment are necessary to provide a complete picture of the learners' knowledge. In developing new forms of assessment, one must determine what methods of assessment are best for evaluating various kinds of knowledge. Both individuals and small groups should be assessed, but for different skills. Small-group learning situations may be useful for measuring the ability to talk about and listen to ideas. Individual assessments might be better for assessing the learner's ability to synthesize knowledge.

Theories of situated cognition, social constructivism, and the influence of the group on an individual's learning can be useful in defining authentic activities and authentic assessment. Although research on situated cognition is still in its infancy, there is evidence that certain activities described by its proponents are similar to those described as worthwhile by mathematics educators. Situated cognition refers to situating learning in the context in which one plans to use the knowledge. Problems must be realistic or authentic in the sense that the applications of knowledge are made apparent to the learner while the learning is taking place, rather than outside of the context in which it could be used (Greeno, 1989).

## SITUATED COGNITION

Situated cognition has developed out of the cognitive apprentice-ship model of instruction (Collins, Brown, & Newman, 1989). The notion of a cognitive apprenticeship comes from traditional ap-prenticeships, where novices learn their trade from a master. The masters share their knowledge with novices, assisting them in developing a skill or product. Similarly, cognitive apprenticeships are designed around the notion that skilled learners can share their knowledge with less skilled learners to accomplish cognitive tasks. Cognitive apprenticeships, however, must model cognitive pro-cesses that are often difficult to externalize so that novices can observe or reflect upon the skills for a particular domain. In theory, the cognitive apprenticeship models offer suggestions for which skills to model for novices, how to provide scaffolding or assistance to less skilled learners, and when to fade such assistance when learners demonstrate they can construct their own meaning. Since the NCTM *Standards* (1989) call for an integration of instruction and assessment, the cognitive apprenticeship model has promise. It provides learners with ways to reflect and correct their perfor-mance based on assessment feedback. This theory does not provide specific guidelines for when and what type of feedback to offer or when to drop back on the amount of assistance provided. If this theory were used to define authentic assessment in an operational way for mathematics knowledge, then such criteria would have to be developed.

Scaffolding or adaptive feedback is important in instruction and assessment. Vygotsky (1978) proposed that assessment con-sider both an individual's actual development or performance on a task without feedback and the potential development or perfor-mance on a task with feedback during test taking. In traditional assessment, where learners' actual development is assessed, it would be difficult to differentiate between two learners who have the same score. The two learners could look quite different from one another if assessed in situations where limited feedback was provided in the test context. Assessment with feedback could measure the learners' potential rather than their actual perfor-mance. Learners may not need feedback the next time they are tested; thus, the test would become a learning experience in and of itself. This is a dynamic and adaptive form of assessment (Frederiksen, 1990; Lajoie & Lesgold, 1992). It is dynamic, because learners can be retested; it is adaptive because learners can learn from the test. Dynamic forms of assessment can provide feedback to learners, giving them ways to improve themselves and opportunities to reach their potential. Tests that serve a learning function may also improve learners' motivation and sense of self-efficacy.

## SOCIAL CONSTRUCTIVISM

The cognitive apprenticeship model is similar to the theory of social constructivism (Vygotsky, 1978) in that learning occurs when one shares knowledge with more capable peers. The NCTM *Standards* (1989) emphasize learners' construction, verification, and revision of mathematical models. They also stress the importance of fostering problem solving, communicating, reasoning, and making connections through small-group or whole classroom discussions. Situated cognition and social constructivist theories fit the NCTM *Standards* well.

Several researchers have examined the construction of mathematical meaning using small groups (Lampert, 1990; Resnick, 1988; Schoenfeld, 1985; Wood, Cobb, & Yackel, 1991). The group helps facilitate reasoning about mathematics and can also foster reflection or use of the metacognitive skills necessary to evaluate mathematical problems (Schoenfeld, 1985). Lampert discusses the importance of finding a common mathematical language for learners to use when communicating ideas. Resnick is particularly clear on the necessity of defining a common core of knowledge to promote the types of dialogue that Lampert refers to in her work. Such dialogues are an important method for demonstrating that mathematical problems may be conducive to multiple, as opposed to single, problem representations. Group problem-solving situations can provide opportunities for discussion prior to implementing mathematical procedures (Resnick, 1988).

The theories reviewed here provide great promise for building authentic activities as well as authentic assessments. There is a gap in the literature on how to operationalize these theories. It is difficult to design groups that will ensure the sharing of cognition and optimize learning for each group member. Because more capable peers assist the less able learners by articulating their cognitive processes, we need to know how to design problem-solving situations that allow for the articulation of such processes, yet provide opportunities for the less skilled to participate in the overall task.

*Authentic Assessment*

Authentic assessment must take place in the context of the learning process. It must consider both the learner and the situation in which the learner is assessed. Authentic assessment must provide information on what the learner knows or does not know and on the developmental changes in such learning. Repeated measures of appropriate learning indicators must be made to obtain a robust picture of the learner's knowledge. These indicators must include a range of cognitive and conative abilities so that multiple perspectives are available for a particular area (see Snow, 1989, for insights regarding the assessment of such learning structures).

Authentic assessment will require instruments that provide in-depth perspectives on learning. Collins, Hawkins, and Frederiksen (1991) have begun to address the best tools for obtaining these perspectives. They suggest that one "picture" does not mean a thousand words when assessing what learners know. At least three different assessment mediums, they suggest, ought to be used to obtain an integrated picture of the learner. The benefits of such mediums as paper and pencil, video, and computers, used jointly, provide a more authentic picture of the learner than a single medium. Paper-and-pencil tests, the traditional forms of assessment, are used to measure students' knowledge of facts, concepts, procedures, problem-solving ability, and text comprehension ability. Collins et al. (1991) suggest broader applications of these tests, using them, for example, to record how students compose texts and documents of various kinds. Students traditionally have been assessed on their essays, but other writing tasks such as letters, reports, memos, drawings, and graphs can also be used to supplement compositions. Paper and pencil can also be used to assess how well students critique the quality of other documents.

Video can be used as a medium for assessing students' communication, explanation, summarization, argumentation, listening, and question-asking and answering skills. Video can also be used to assess student interactions in the context of cooperative problem-solving activities. Video records of dynamic interactions can be scored at a later time. They provide opportunities for scoring oral presentations, explanations provided in a small group setting, and joint problem-solving activities.

The computer can provide a further perspective on the learner. It can effectively track the process of learning as well as a learner's response to feedback. It can also simulate realistic situations in the classroom. The computer provides opportunities for assessing the dynamic nature of problem solving and opportunities to systematically vary the instructional environment on the feedback dimension and observe the effects on learning outcomes (Lajoie & Lesgold, 1992). The feedback dimension offers us a novel mechanism for assessing how well or how poorly individuals respond to certain learning environments. The ability to track student performance provides opportunities for assessing such strategic aspects of knowledge as hypothesis formation and hypothesis verification or for assessing motivational aspects of learning—how persistent students are at trying to solve the problem—as well as actual learning outcomes. Thus, computers make possible the dynamic assessment of relevant criteria. Computers can also provide a less structured means of assessment, in which students are tracked as they explore mathematical content area (Shaughnessy, 1992).

Collins et al. (1991) suggest that the use of these three mediums of assessment will provide a more robust picture of the learner. The assessment medium, however, is only as authentic as the task that the learner is being tested on. Care must be taken to define the types of student records that will be collected and to ensure that such records reflect the performance indexes most relevant to that medium.

Finally, the purpose or use of assessment must be considered. If assessment results are used by the learners or teachers, then the assessment tools must be available in the classroom on a regular basis, weaving together instruction and assessment. The interdependence of instruction and assessment has been referred to as a *systemic approach* (Frederiksen & Collins, 1989; Salomon, 1991) and often used in the context of performance assessment (Baron, 1990; Linn, Baker, & Dunbar, 1991; Wolf, Bixby, Glenn, & Gardener, 1991). Learners should be able to use the tests as tools to reflect on their strengths and weaknesses (Nitko, 1989). Tests or assessment tools should be transparent in the sense that those who are being assessed understand the criteria on which they are being judged so they can improve their performance (Frederiksen & Collins, 1989). Frederiksen and Collins suggest that one way to ensure that the assessment criteria are transparent is to provide a library of exemplars for students to visit. This library provides copies of records of student performances that have been critiqued by master assessors in terms of the relevant criteria. Such a library would help students evaluate their own performance and perhaps provide landmarks of success for which to strive. In addition to self-assessment, feedback should be given to students after a test is taken to help them improve their performance. Teachers can be assisted in using the assessment tools to determine what concepts students have misinterpreted.

## PRINCIPLES FOR OPERATIONALIZING AUTHENTIC ASSESSMENT

We seek to define and operationalize authentic assessment to improve learning. Thus, students should find undertaking an assessment task a learning experience. And teachers should learn what their students know or do not know as a result of the assessment task. Some tentative principles for operationally defining authentic assessment grow out of the theories and literature reviewed:

1. It must provide us with multiple indicators of the learning of the individual in those cognitive and conative dimensions that affect learning. The cognitive dimensions should include content knowledge, how that knowledge is structured, and how information is processed with that knowledge. The conative dimensions should address students' interest in and persistence on tasks as well as their beliefs about their ability to perform. Student interest in a topic often increases in conjunction with a deeper conceptual knowledge of that topic. Students' choices may reflect their level of engagement and interest. These indicators must be examined repeatedly if they are to provide us with information on learning transitions or developmental maturity. Multiple mediums of assessment are necessary to provide valid indicators; that is, indicators that we define as authentic. One measure, obtained by a single medium, is unlikely to provide us with sufficient information on an individual. Varied types of procedures are necessary for gathering assessment information (Collins et al., 1991; Romberg, 1992).

2. Authentic assessment must be relevant, meaningful, and realistic. It must be instructionally relevant, as indicated by its alignment with the NCTM *Standards* (1989). It must relate to pure and applied tasks that are meaningful to students and that provide them with opportunities to reflect, organize, model, represent, and argue within and across mathematical domains.

3. It must be accompanied by scoring and scaling procedures constructed in ways appropriate to the assessment tasks.

4. It must be evaluated in terms of whether it improves instruction, is aligned with the NCTM *Standards*, and provides information on what the student knows.

5. It must consider racial or ethnic and cultural biases, gender issues, and aptitude biases.

6. It must be an integral part of the classroom. Because teachers are more likely to teach the information to students that appears on tests, assessment tasks must be aligned with authentic activities such as those outlined in the NCTM *Standards*. Teachers need to be an integral part of the assessment loop so that they can learn from assessment information and structure their instruction accordingly.

7. It must consider ways to differentiate between individual and group measures of growth and provide for ways of assessing individual growth within a group activity.

Alternatives to paper-and-pencil multiple-choice tests do exist. Those listed here incorporate several principles of, and hold promise as authentic testing forms for, the assessment of mathematics learning:

*Australian IMPACT Project*

A set of studies was conducted in Australia to facilitate communication within the college-level mathematics classroom (Clarke, Stephens, & Waywood, 1992). Journals were kept by students and used by both teachers and students to foster a dialogue about what the students were learning. The quality of student journals progressed from simple narratives that described concepts, to summaries that integrated mathematics knowledge, to dialogues regarding what questions should be addressed and what meaning could be constructed as well as the connections of their work with other mathematics knowledge. These journals were beneficial to both teachers and students because they provided opportunities for dialogue not possible during a regular class period. They demonstrate that instruction and assessment can be integrated in the classroom. Student journals could provide us with new techniques for authentically assessing mathematical communication skills by providing the mechanism for examining transitions in developmental maturity in these skills.

*Vermont Portfolios*

Portfolios are promising as an assessment tool because they provide multiple examples of the students' work and provide students with experience in generating mathematical ideas, seeing mathematics as part of the culture, and being enculturated in the mathematics experience. What is particularly intriguing about portfolios is that multiple audiences can use them to obtain knowledge of the learners, teachers, and curriculum. Guidelines are needed, however, on how to score such materials.

### Connecticut Common Core of Learning Project

The Connecticut Common Core of Learning Project (Baron, Forgione, Rindone, Kruglenski, & Davey, 1989) provides learners with authentic uses of mathematics. Assessment consists of in-depth evaluations of learners in the context of problem-solving situations that may take a week to complete. This project embodies systemic assessment in that instruction and assessment are integrated. Teachers are provided with assessment tools, in the form of scoring templates, that facilitate their task of assessing learning. This project provides an example of how to examine both individual and cooperative group problem-solving activities and, in doing so, provides insights as to how students form their own hypotheses by comparing theirs with other hypotheses and how they generalize concepts from one problem situation to another.

### The California Assessment Program

The California Assessment Program (1989) has addressed the concerns of the NCTM *Standards* (1989) by providing students with opportunities to demonstrate their construction of mathematical meaning consistent with their mathematical development. Open-ended questions are provided that give students opportunities to think for themselves and express their ideas. Communication is fostered in classroom discussions as well as in writing tasks. The data from this project provides a wealth of information regarding students' misconceptions and reasoning abilities.

### Cognitively Guided Instruction

In the Cognitively Guided Instruction project (Carpenter, Fennema, Peterson, & Carey, 1987; Carpenter & Fennema, 1988), instructional decisions are based on careful analyses of student knowledge and the goals of instruction. Problems are selected that closely match the student's knowledge level. The assessment emphasis is on the learning processes of students. Individual and group data are collected.

*Problem Situations*

De Lange (1987) has designed mathematical problem situations composed of multiple items with varied levels of difficulty. In his assessment of the Hewet Mathematics Project in The Netherlands, five different tasks were used to gather information: a timed written task, two-stage tasks, a take-home examination, an essay task, and an oral task. These provided a multifaceted evaluation of the learner. The two-stage tasks are especially interesting, in light of our principles of authentic assessment. Stage one includes open-ended questions and essay questions. These items are scored and returned to the student. In stage two, students are provided with their scores from stage one, allowed to take the stage-one tests home, and given as long as three weeks to answer the same questions. The final assessment includes scores from stage one and stage two. Students can learn from their mistakes and from the feedback regarding their mistakes, making the testing process an interactive one that assists students in reaching their potentials.

*Superitems*

Superitems are designed to elicit mathematical reasoning about mathematical concepts (Collis & Romberg, 1991). The items are built to assess four different levels of mathematical maturity. At level four, the most mature level, the learner must articulate some understanding of the mathematical concepts either in words or symbols. The tasks can be used to obtain measures of developmental reasoning and to serve as a first step in the identification of learning transitions in mathematical content areas.

Many other alternative forms of assessment demonstrate a promise of being authentic. For instance, there are software programs and multimedia approaches to learning that allow learners to explore multiple forms of representations when learning mathematics (one example is the function analyzer, which provides visual representations of mathematical concepts [Harvey, Schwartz, & Yerushalmy, 1988]). We need to evaluate these new technologies along with the alternatives listed previously and consider what other options exist for extending our notion of authentic assessment.

I have laid out a tentative framework for the development of authentic forms of assessment. These, and other alternative forms of assessment that incorporate new technologies, hold promise for fitting within an operational definition of *authentic assessment*.

Several parts of the framework require additional research. We will need to determine how cognitive and conative learning indicators can be operationalized in the context of an assessment task. We will need to study how to obtain frequent and valid measures of learners' performances. And we will need to define what we are assessing in individual and group situations. Finally, when considering the multiple audiences that may use measures obtained by authentic means, we must keep equity issues in focus.

### REFERENCES

Ballew, H. (1981). "Applications of statistics and probability to genetics." In A. P. Shulte & J. R. Smart (eds.), *Teaching statistics and probability*, pp.173–181. Reston, VA: National Council of Teachers of Mathematics.

Baron, J. B. (1990). "Performance assessment: Blurring the edges among assessment, curriculum, and instruction." In A. B. Champagne, B. E. Lovitts, & B. J. Calinger (eds.), *Assessment in the service of instruction*, pp. 127–148. Washington DC: American Association for the Advancement of Science.

———, Forgione, P., Rindone, D., Kruglenski, H., & Davey, B. (1989). "Toward a new generation of student outcome measures: Connecticut's Common Core of Learning Assessment." Paper presented at the annual meeting of the American Educational Research Association, San Francisco.

Bertin, J. (1973). *Semiologie Graphique*, 2nd ed. The Hague: Mouton-Gauthier. English translation by W. Berg & H. Wainer as *Semiology of graphics*. Madison: University of Wisconsin Press, 1983.

Bransford, J., Hasselbring, T., Barron, B., Kulewicz, S., Littlefield, J., & Goin, L. (1988). "Uses of macro-contexts to facilitate mathematical thinking." In R. Charles & E. Silver (eds.), *The teaching and assessing of mathematical problem solving*. Hillsdale, NJ: Lawrence Erlbaum Associates.

Bryant, P. (1984). "Geometry, statistics, probability: Variations on a common theme." *The American Statistician* 38, no. 1:38–48.

California Assessment Program. (1989). *A question of thinking: A first look at students' performance on open-ended questions in mathematics*, Technical Report, CAP. Sacramento, CA: California State Department of Education.

Carpenter, T. P., & Fennema, E. (1988). *Research and cognitively guided instruction*. Madison, WI: National Center for Research in Mathematical Sciences Education, Wisconsin Center for Education Research.

Carpenter, T. P., Fennema, E., Peterson, P. L., & Carey, D. A. (1987). "Teachers' pedagogical content knowledge in mathematics." Paper presented at the American Educational Research Association, Washington, DC.

Clarke, D., Stephens, M., & Waywood, A. (1992). "Communication and the learning of mathematics." In T. Romberg (ed.), *Mathematics assessment and evaluation: Imperatives for mathematics educators*. Albany: SUNY Press.

Collins, A., Brown. J. S., & Newman, S. (1989). "Cognitive apprenticeship: Teaching the craft of reading, writing, and mathematics." In L. B. Resnick (ed.), *Knowing, learning, and instruction: Essays in honor of Robert Glaser*, pp. 453–494. Hillsdale, NJ: Lawrence Erlbaum Associates.

Collins, A., Hawkins, J., & Frederiksen, J.R. (1991). *Three different views of students: The role of technology in assessing student performance*, Technical Report #12. New York: Center for Technology in Education, April.

Collis, K. F., & Romberg, T. A. (1991). "Assessment of mathematical performance: An analysis of open-ended test items." In M. C. Wittrock & E. L. Baker (eds.), *Testing and cognition*, pp. 82–130. Englewood Cliffs, NJ: Prentice-Hall.

de Lange, J. (1987). *Mathematics insight and meaning*. Utrecht: OW & OC.

Fong, G. T., Krantz, D. H., & Nisbett, R. E. (1986). "The effects of statistical training on thinking about everyday problems." *Cognitive Psychology* 18:253–292.

Frederiksen, N. (1990). "Introduction." In N. Frederiksen, R. Glaser, A. Lesgold, & M. Shafto (eds.), *Diagnostic monitoring of skill and knowledge acquisition*, pp. ix–xvii. Hillsdale, NJ: Lawrence Erlbaum Associates.

Frederiksen, J. R., & Collins, A. (1989) "A systems approach to educational testing." *Educational Researcher* 18, no. 9:27–32.

Greeno, J. (1989). "A perspective on thinking." *American Psychologist* 44:134–141.

Harvey, J. G., Schwartz, J., & Yerushalmy, M. (1988). *Visualizing algebra: The function analyzer* (computer software). Pleasantville, NY: Sunburst.

Lajoie, S. P., & Lesgold, A. (1992). "Dynamic assessment of proficiency for solving procedural knowledge tasks." *Educational Psychologist* 27, no. 3:365–384.

Lampert, M. (1990). "When the problem is not the question and the solution is not the answer: Mathematical knowing and teaching." *American Educational Research Journal* 27, no. 1:29–63.

Linn, R. L., Baker, E. L., & Dunbar, S. B. (1991). "Complex, performance-based assessment: Expectations and validation criteria." *Educational Researcher* 20, no. 8:15–21.

Mosteller, F. (1988). "Broadening the scope of statistics and statistical education." *The American Statistician* 42, no. 2:93–99.

National Council of Teachers of Mathematics (NCTM) (1989). *Curriculum and evaluation standards for school mathematics*. Reston, VA: Author.

Nitko, A. J. (1989). "Designing tests that are integrated with instruction." In R. L. Linn (ed.), *Educational Measurement*, 3rd ed., pp. 447–474. Washington, DC: Macmillan.

Pereira-Mendoza, L., & Swift, J. (1981). "Why teach statistics and probability—a rationale." In A. P. Shulte & J. R. Smart (eds.), *Teaching Statistics and Probability Yearbook*, pp. 1–7. Reston, VA: Author.

Pollak, H. (1987, May). Notes from a talk given at the Mathematical Sciences Education Board Frameworks Conference. Minneapolis, MN.

Reed, S. K. (1985). "Effects of computer graphics on improving estimates to algebra word problems." *Journal of Educational Psychology* 77, no. 3:285–298.

Resnick, L. B. (1987). "Learning in school and out." *Educational Researcher* 16:13–20.

———. (1988). "Teaching mathematics as an ill-structured discipline." In R. Charles & E. A. Silver (eds.), *The teaching and assessing of mathematical problem solving*, pp. 32–60. Reston, VA: National Council of Teachers of Mathematics.

Romberg, T. (1992). "Evaluation: A coat of many colors." In T. Romberg (ed.), *Mathematics assessment and evaluation: Imperatives for mathematics educators*. Albany: SUNY Press.

Rosenheck, M. (1991). "The effects of instruction using a computer tool with multiple, dynamically, and reversibly linked representations on students' understanding of kinematics and graphing." Unpublished doctoral dissertation, University of Wisconsin-Madison.

Salomon, G. (1991). "Transcending the qualitative-quantitative debate: The analytic and systemic approaches to educational research." *Educational Researcher* 20, no. 6:10–18.

Schoenfeld, A. (1985). *Mathematical problem solving*. New York: Academic Press.

Shaughnessy, J. M. (1992). "Research in probability and statistics: Reflections and directions." In D. Grouws (ed.), *Handbook for research in mathematics education*, pp. 465–494. New York: Macmillan.

Singer, J. D., & Willett, J. B. (1990). "Improving the teaching of applied statistics: Putting the data back into data analysis." *The American Statistician* 44, no. 3:223–230.

Snow, R. E. (1989). "Toward assessment of cognitive and conative structures in learning." *Educational Researcher* 18, no. 9:8–14.

Tanner, M. A. (1985). "The use of investigations in the introductory statistics course." *The American Statistician* 39, no. 4:306–310.

Vygotsky, L. S. (1978). *Mind in society: The development of higher psychological processes.* Cambridge, MA: Harvard University Press.

Wainer, H. (1992). "Understanding graphs and tables." *Educational Researcher* 21, no. 1:14–23.

Whetzel, D. (1992). *The Secretary of Labor's Commission on Achieving Necessary Skills.* (Report No. EDO-TM-92-1). Washington, DC: Office of Educational Research and Improvement (ERIC Document Reproduction Service No. ED339749).

Wolf, D., Bixby, J., Glenn III, J., & Gardener, H. (1991). "To use their minds well: Investigating new forms of student assessment." *Review of Research in Education* 17:31–74.

Wood, T., Cobb, P., & Yackel, E. (1991). "Change in teaching mathematics: A case study." *American Educational Research Journal,* 28, no. 3:587–616.

Zarinnia, E. A., & Romberg, T. (1992). "A framework for the California Assessment Program to report students' achievement in mathematics." In T. A. Romberg (ed.), *Mathematics assessment and evaluation: Imperatives for mathematics educators,* pp. 242–284. Albany: SUNY Press.

# 3 ❖ Sources of Assessment Information for Instructional Guidance in Mathematics[1]

*Edward A. Silver and
Patricia Ann Kenney*

Mathematics education reform is currently a topic of great interest in this country, with much of the attention focused on new goals for the school mathematics curriculum and instruction. At the heart of the reform discussion and fueled by evidence of poor student performance in mathematics on national and international surveys is a call for raising national standards to a level considered to be "world class." The reform movement and the perceived need for national standards of mathematics achievement have led members of the business, government, and education communities to focus attention on assessment, especially on the development and implementation of a national examination system to monitor progress toward the attainment of higher performance standards. Far less attention, however, is being paid to discussions about the ways in which assessment information—whether from national or local tests or from classroom assessments—can be used to guide instructional decision making to improve mathematics teaching and learning for all students. This topic—assessment for the purposes of instructional guidance—is the focus of this chapter.

Education professionals—teachers, school administrators, curriculum supervisors, counselors, and curriculum developers—routinely make decisions that affect the form and content of students' instructional experiences in mathematics. For example, school administrators determine the allocation of resources to support instructional programs; teachers and counselors make decisions regarding student access to portions of the instructional program, such as special classes for exceptional students; and curriculum developers and supervisors make decisions that influence the content of instruction and, often through staff development, the form in which the content will be taught. These macro-level decisions can be and often are informed by student assessment information, as are the micro-level decisions frequently made in classrooms by teachers.

To guide the instructional programs they provide to students, classroom teachers make frequent decisions about the differentiation of instruction, about the inclusion of topics in a lesson sequence or homework assignments, about the pacing of the coverage of topics, and about the selection of teaching methods. Their decisions are influenced by information obtained from formal and informal assessments of their students. For example, a mathematics teacher who plans a unit of instruction on the topic of measurement for her seventh-grade students might consider information available from a wide variety of sources:

- Data from state-level or district-level tests that suggest areas of strength or weakness in her students' knowledge of measurement;
- Information from a brief diagnostic test that could be designed and administered before beginning the unit;
- Data on her students' performance in prior units on related topics (e.g., geometry, rational number computation) or on prior units in which students have been particularly successful;
- Records of previous students' final performance and achievement on the measurement unit taught in recent years;
- Observations of students' level of engagement with various forms of instructional activities.

Each of these sources of information could influence decisions about the content of the unit (e.g., selection of topics and worthwhile mathematical tasks) and the method of teaching (e.g., modifying a problem-based approach that was particularly successful in a unit on statistics or adapting forms of discourse that have enhanced student learning of other topics), as well as decisions about pacing and possible differentiation of instruction. Moreover, these and other information sources embedded in ongoing instructional activities (e.g., observing students working in small groups, talking with students, grading homework, or evaluating extended projects) could provide additional information during the teaching of the unit that can lead to modifications to the original plans, such as adjusting the pacing or selecting additional or alternative mathematical tasks.

The measurement unit example showcases many of the forms that assessment can take, highlighting the fact that although assessment is often thought of as synonymous with paper-and-pencil testing, it actually includes techniques that collect a full range of information about students and the classroom environment. The information obtained from this broad array of assessment activities provides many sources that teachers and others can use when making instructional decisions.

In this chapter, we examine a variety of sources of assessment information available to teachers and other educational professionals as they make important instructional decisions. First we consider the information available from external sources, such as standardized achievement tests and international or national assessment surveys. Then, we turn our attention to many kinds of internal, classroom-based assessments that can provide important information on which instructional decisions could be based. The array of internal assessments to be considered includes not only traditional assessments such as classroom tests and homework but also periodic observations, questioning, and performance on projects and open-ended tasks that provide opportunities for students to demonstrate the nature and extent of their understandings and proficiencies.

## EXTERNAL ASSESSMENTS AS SOURCES OF INFORMATION FOR INSTRUCTIONAL GUIDANCE: POSSIBILITIES AND LIMITATIONS

Using instructional time to administer some kind of externally mandated or externally developed assessment is a quite common occurrence in most mathematics classrooms in the United States. School districts or state departments of education frequently require students to take mathematics tests in order to use the test results for program evaluation or student placement, diagnosis, or summative evaluation.[2] In addition to these externally mandated tests, some schools or districts also volunteer to participate in external assessments that have as their main goal monitoring or comparison of student proficiency and achievement on a national or international level. After these external tests are administered—sometimes long after—the results become available to administrators and classroom teachers, who are then left with the task of interpreting and evaluating the information and making instructional decisions.

Given the pervasive presence of externally mandated tests in the lives of mathematics teachers and students, it is important to consider ways in which the results of these tests may or may not provide information that could be in some way useful to classroom teachers or to others for instructional guidance. This section of the chapter discusses external assessments of two different types, standardized tests and large-scale surveys, and provides some examples of ways in which the results of these tests might or might not assist mathematics teachers and other education professionals in making instructional decisions.

### Standardized Tests

At some time during the school year every teacher must relinquish class time to administer a battery of standardized tests. These tests are regularly and widely used to provide a means of measuring individual student achievement. The term *standardized* refers to the fact that these tests are designed to be administered, scored, and interpreted in the same way each time they are used. Many standardized tests are nationally normed to allow for comparisons among students; others are developed to compare individual student achievement on a predetermined set of content objectives. Items on standardized tests are typically multiple choice, a format that provides for highly efficient measurement and ease of scoring. Typically, a mathematics test in a commercially published test battery (e.g., the California Achievement Test, the Iowa Test of Basic Skills, and the Stanford Achievement Test) consists of two or three parts: computation, concepts, and applications or problem solving.

Most school districts rely on multiple-choice tests developed by commercial publishers to provide mathematics achievement information for their students. Many states and some school districts have developed and administer regularly other standardized tests that are tied more closely to their curriculum objectives. These tests can be used for a variety of purposes: to evaluate the success of a school's instructional program in achieving stated objectives, to certify student competence for high school graduation, to identify students in need of remedial attention, and so on. Because commercial standardized tests and those developed by states or districts may serve differentially as sources of assessment information for instructional guidance, they will be treated separately.

*Commercial Standardized Tests.*  In general, publishers of commercial standardized tests declare their purpose to be the improvement of instruction and learning. For example, one test publisher states that "the most important use of achievement test results is to help improve student learning through instruction" (Science

Research Associates, 1979, p. 32). Nevertheless, these tests are at the heart of the criticism and calls for change in current testing practice, and they have been involved in many skirmishes in the battle for education reform in the United States.[3]

One source of the conflict over the use of commercial standardized tests as measures of mathematics achievement is the perceived mismatch between the vision of mathematical proficiency and competence proposed in publications of the National Council of Teachers of Mathematics (1989) or the National Research Council (1989) and the definition implied by the content of such tests. In its report on the future of mathematics education in the United States, *Everybody Counts*, the National Research Council addressed this mismatch: "As we need standards for curricula, so we need standards for assessment. We must ensure that tests measure what is of value, not just what is easy to test. If we want students to investigate, explore, and discover, assessment must not measure just mimicry mathematics" (1989, p. 70). Critics have argued that the content of current standardized tests stands in opposition to the reform vision of competence and proficiency, in which such themes as thinking, reasoning, complex performance, and problem solving are emphasized in addition to or in place of knowledge and basic skill performance. Commenting on the failure of current tests to serve as appropriate symbols of an authentic vision of mathematical proficiency, Silver and Kilpatrick (1988) argued: "Another function of testing is to signal to students, teachers, and the general public those aspects of learning that are valued. When students ask, 'Is that going to be on the test?' they are inquiring as to the value of the knowledge in question. In general, current tests place greater emphasis on those aspects of the curriculum that are relatively easy to assess than on those aspects that are highly valued by professionals in the field of mathematics education" (p. 180).

A closely related criticism of commercial standardized tests is the narrow range of curriculum content that typically is covered. Deciding the content of these tests is a consensual, market-driven process not unlike that associated with creating textbooks (Tyson-Bernstein, 1988). To produce a test that is marketable throughout the country, test publishers need to ensure that its content at each grade level resides at the intersection of the curricular goals of the states publishing such goals. Naturally, this leads to narrowing the range of content that might be included at any given grade level, thereby resulting in an overemphasis in mathematics on basic computational skills and the virtual exclusion of items that measure higher level thinking, reasoning, and problem solving (California Mathematics Council, 1986; Romberg, Wilson, & Khaketla, 1989).

In addition to the misalignment of test content and curricular goals, another frequent criticism of standardized tests involves the excessive use of the multiple-choice answer format. Because of the demands of testing large numbers of students in short amounts of time, commercial test developers have made almost exclusive use of multiple-choice items. This format does not allow questions in which students are required to produce their own answers, display the processes used to obtain an answer, explain the thinking or reasoning associated with their responses, or exhibit alternative approaches to or interpretations of a problematic situation. Moreover, the use of multiple-choice formats is usually associated with the imposition of a severe time limitation, which prevents students from displaying their competency under conditions that are more felicitous for optimal performance.

The often fiery criticism of commercial standardized tests and other forms of externally mandated testing has been fueled not only by these limitations but also by evidence that the widespread use of these tests can limit and negatively affect the quality of mathematics instruction. Some researchers (Romberg, Zarinnia, & Williams, 1989; Smith, 1991) have suggested that teachers are influenced by their perceptions of the content of externally mandated tests, especially when the test results are viewed as having important consequences for them or their students. In particular, the research suggests that teachers tend to narrow their instruction by giving a disproportionate amount of their time and attention to teaching the specific content most heavily tested, rather than teaching underlying concepts or overarching principles, or rather than teaching untested or less tested areas (e.g., geometry, statistics) that are also expected to be part of the curriculum. Other studies (e.g., LeMahieu & Leinhardt, 1985) have found that the role of the teacher as instructional decision maker is often influenced by the perceived content of commercial standardized or other externally mandated tests. In this way, many critics charge that the widespread use of multiple-choice testing contributes to a "dumbing down" of instruction, in which skills are taught only in the form required for the test rather than for more realistic or natural applications (Darling-Hammond & Wise, 1985).

Beyond the influence that "teaching to the test" may have in shaping instructional practice, it can also diminish the value of the information obtained from testing. As Shepard has noted, "The more we focus on raising test scores, the more instruction is distorted, and the less credible are the scores themselves" (1989, p. 9). Rather than serving as accurate indicators of student knowledge and performance, the tests become indicators of the amount of instructional time and attention paid to the narrow range of skills and competencies assessed.

It should come as no surprise that these criticisms and limitations of commercial standardized tests decrease the usefulness of the information obtained from such tests for instructional guidance. At best, commercial standardized mathematics achievement tests can provide teachers with some general within-student information (a student's mathematics achievement in comparison with his or her achievement in other subject areas) and across-student comparisons of mathematics achievement on the tested content. Within a school or district, teachers, administrators, and counselors might use several years of test results as a crude indicator of improvement or decline in performance over time. Such monitoring might detect gross changes that could be useful in directing instructional attention at needed targets. For example, a pattern of very low performance on certain subsections of a test could suggest areas that are not receiving adequate amounts of instructional attention. Armed with this information, administrators and teachers could decide whether or not the importance of the mathematics content in those test sections warrants a change in the allocation of instructional resources. Nevertheless, because of the limited scope of the content and item formats on these tests, it is unlikely that the information produced by commercial standardized tests will be very helpful in providing detailed instructional guidance for teachers, administrators, or supervisors trying to move mathematics programs in the direction of the National Council of Teachers of Mathematics' (1989) *Curriculum and Evaluation Standards*.

One area in which standardized test information is often used for instructional guidance is in student placement. For example, the results of commercial standardized tests have long been used to determine student eligibility for placement in special programs such as grade 8 algebra or Chapter I remedial instruction. In the case of Chapter I, the tradition has been to use results of commercial, norm-referenced standardized tests as the sole criterion for placement. Recent changes in federal laws, however, now allow school personnel to use other methods of assessment, such as performance tasks and interviews (Stenmark, 1989). These new regulations should allow administrators, counselors, and teachers responsible for making such placement decisions to supplement the limited information available from commercial standardized tests and to use multiple sources of assessment information to make the best instructional decisions for the target students.

*State-Level and District-Level Tests.* In addition to providing class time for the administration of commercial standardized tests, an increasing number of teachers are finding that additional time must be provided for the administration of state-level and district-level assessments. In response to actual or perceived demands for public accountability, many states and school districts have developed their own tests for many purposes, including monitoring individual student progress, evaluating program effectiveness, identifying students in need of remedial assistance, or certifying competence for high school graduation. Because of the difficulty of constructing good assessment measures, these tests have often been modeled after commercial standardized tests, with an exclusive use of multiple-choice formats. Furthermore, commercial test publishers are frequently called upon to develop state-level and district-level achievement tests. Hence, these local tests have a high probability of inheriting the limitations of commercial standardized tests.

State and local standardized tests, however, usually differ from commercial standardized tests in one important aspect. In contrast to commercial standardized tests, which must be developed without reference to any particular set of curriculum objectives, state and local tests can be designed to reflect a specific set of subject-matter objectives. Because of this ability to relate test items to particular objectives, many state and local tests are developed as criterion-referenced tests. In contrast to the commercial standardized, norm-referenced tests, for which a student's performance is compared to that of a national comparison group, student performance on criterion-referenced state or local tests can be described in reference to a particular objective or set of objectives.

Although all tests involve students responding to a set of questions, the analysis and use of the information obtained from the testing is somewhat different in the various types of state-level and district-level testing programs. In some cases, information is gathered to report at a classroom, school, or school district level the proficiency of the student population with respect to specific objectives. Results from these tests are usually reported in terms of the percentage of students (in the class, school, or district) who scored at a specified level. For example, if the passing level for measurement concepts is set at the 70 percent level (i.e., a student correctly answers 70 percent of the items testing that objective), then the results will be reported in terms of the percentage of students who achieved that level. In some cases the test may be designed so that there are different forms, each of which contains only a few of the questions for a particular subject at a particular grade level. Because different students may not answer the same or comparable questions, it will be impossible to derive from such a test information about relative performance at the student level; hence, program-level reporting is preferable.

Results from these program-centered tests can be of value in the instructional planning of a classroom teacher, especially because the state- or district-developed set of test objectives is likely to be closer to those that constitute the curriculum in that individual teacher's classroom. For example, a teacher can observe that students in his or her district are performing well on items that involve the solution of linear measurement problems, but that these same students are not doing as well on items that involve area measure. Knowing that area measurement is a concept that may be difficult for the students in that classroom, the teacher can consider adjusting the pacing of the unit by increasing the time spent on area within the measurement unit or adjusting the teaching method to one that uses more concrete examples to build a conceptual bridge from linear measure to area measure. Curriculum developers and mathematics supervisors can also use information from state- and district-level tests to influence curriculum revision and staff development activities.

For some state- or district-level tests, results are reported by individual student and used to make basic educational and instructional decisions, usually related to students whose scores do or do not exceed some predetermined "cut score." For example, minimum-competency tests are often used for placement into remedial instruction programs; passing a "graduation test" often determines whether a student earns a high school diploma; some states even have tests that a student must pass to be promoted to the next grade (Airasian, 1991). Results from these tests obviously provide information upon which instructional decisions are made—students are assigned to remedial programs or required to "repeat" a course of instruction. At the classroom level, the tests may also provide some information for the teacher. For example, knowing an individual student's strengths and weaknesses on a minimum-competency test could provide information a teacher could use to assign extra work or adjust instructional approaches or assign a student to a cooperative learning group. Aggregated information about group performance on certain topic areas may also provide some information that could inform decisions about the allocation of time to different units of instruction or the pacing of lessons. However, it is important to remember that the value of the information to educational professionals who are trying to move classroom instruction in the direction of mathematics education reform depends heavily on the relationship between the reform vision and the actual test content and format. If the test content reflects a narrow conception of mathematics or if the test format samples narrowly from the wide range of mathematical performances, it is unlikely that such a test can provide useful information for instructional guidance.

The mismatch of test content and a specific vision of mathematical proficiency becomes even more problematic because of the ways in which test scores are often used. It is not at all uncommon for the scores to be used as the basis for comparisons among schools. In fact, lists of schools and their average student performance are often published on the pages of the local newspaper, thereby inviting comparisons between high-scoring schools and low-scoring schools, without regard for the many other factors that would have to be taken into account in making valid comparisons. An even more dangerous, and unfair, practice is the use of such test scores for the evaluation of individual teachers within schools, a use for which the scores were never intended. As a means of obtaining information for use in making instructional decisions, these practices are dangerous because they can lead administrators to pressure teachers to shape their instruction toward the test content and test format, as discussed earlier in the context of commercial standardized tests. The words of one teacher sum up these concerns: "My principal puts a great deal of emphasis on our school's performance on the state-mandated basic skills test. He's very concerned about how we do compared to neighboring schools when the results are published in the local paper. The emphasis is to include in instruction topics on the test but [that are] not yet in our curriculum and to give tested topics more instructional time" (Airasian, 1991, p. 361).

For teachers and other decision makers in states or districts that use tests based on appropriate mathematics content and that report and use scores fairly, these tests can provide some information for the purposes of instructional guidance. Looking beyond state- and district-level boundaries, mathematics teachers may also gain some helpful information from national and international surveys of students' mathematical achievement.

### National and International Assessments

Recent proclamations (e.g., *Goals 2000*) concerning the dire state of students' mathematics achievement in the United States and the need for massive improvement to reach "world class standards" of mathematics performance have been stimulated by data obtained from national and international surveys of mathematical knowledge. Much of the attention given to this issue has focused on the need to create national standards for mathematics proficiency and a national testing system to measure students' attainments. If and when a national testing system is designed and implemented, the data generated from its implementation may be useful for instructional guidance in some ways, especially if the assessment tasks used in the test embody the reform vision of mathematical proficiency. At this time, however, education professionals seeking to use such information for instructional guidance are limited to data available from several existing surveys of mathematics knowledge. Although these surveys have had some impact in mobilizing public concern about mathematics achievement, the remainder of this section deals with ways in which information generated in such surveys might be useful to those making instructional decisions.

*International Surveys.*  Beyond sounding alarms because of the poor performance of U.S. students compared to their counterparts in other countries—alarms that some critics (e.g., Husen, 1983; Rotberg, 1990, 1991) think may be inappropriate and unnecessary—international comparisons offer useful benchmarks against which to gauge the performance of students in this country. The information available from international assessment surveys is certainly interesting, yet its direct relevance to the daily instructional decisions made by classroom mathematics teachers is probably minimal. Due to design constraints of an assessment that has to consider students at a variety of grade levels from many nations, the information may be far removed from the content or focus of any particular teacher's instructional program.

Despite these limitations, and whether or not their own students participate directly in the testing, teachers and other instructional decision makers can get a global view of student knowledge and performance in mathematics and may be able to glean some information useful in instructional guidance from these international performance surveys. For example, reports of the results of international surveys such as the Second International Mathematics Study (SIMS) (McKnight et al., 1987) and the International Assessment of Educational Progress (IAEP) (Lapointe, Mead, & Phillips, 1989) contain sample items and further information and commentary that might help educators address instructional issues related to student achievement in various content areas (e.g., algebra, geometry, probability, and statistics) and in specific ability areas (e.g., problem solving).

In the IAEP, sample items were linked to levels of mathematics proficiency, as shown in figure 3.1. One way in which middle school mathematics teachers might use the IAEP proficiency levels and sample items would involve having students respond to the sample items and then looking at student performance with respect to five proficiency levels. Another option might be for a group of teachers to meet and examine the content categories included in international assessments and then compare them to content categories that form the local mathematics curriculum.

Beyond these opportunities for teachers to use specific results or performance summaries, it is also possible that teachers or other instructional decision makers would be able to use information related to more general trends or explanations that emerge from such surveys. For example, based on a review of the results from an international survey, a mathematics supervisor might make an impact on instruction by planning staff development sessions on certain relevant findings, or a curriculum developer might modify some exciting curricular units. Moreover, based on consideration of the general findings from SIMS to the effect that the mathematics curriculum in the United States lacked focus and depth (McKnight et al., 1987), district or state mathematics supervisors might engage teachers in efforts aimed at curriculum development or topical rearrangement to address the problem.

*National Assessments.* The National Assessment of Educational Progress (NAEP) is a general source of mathematics achievement information that may have direct relevance to classroom teachers, curriculum developers, mathematics supervisors, and other mathematics instruction decision makers in the United States. There have been six such national assessments in mathematics, with the results from the last NAEP released in April 1993. The purpose of the NAEP mathematics assessment is to provide a general picture of what students "know and can do" in

LEVEL **Perform Simple Addition and Subtraction**

**300** Students at this level can add and subtract two-digit numbers without regrouping and solve simple number sentences involving these operations.

LEVEL **Use Basic Operations
to Solve Simple Problems**

**400**

Students at this level can select appropriate basic operations (addition, subtraction, multiplication, and division) needed to solve simple one-step problems. They are capable of evaluating simple expressions by substitution and solving number sentences. They can locate numbers on a number line and understand the most basic concepts of logic, percent, estimation, and geometry.

LEVEL **Use Intermediate Level Mathematics Skills to
Solve Two-Step Problems**

**500**

Students at this level show growth in all mathematics topics in the assessment. They demonstrate an understanding of the concept of order, place value, and the meaning of remainder in division; they know some properties of odd and even numbers and of zero; and they can apply elementary concepts of ratio and proportion. They can use negative and decimal numbers; make simple conversions involving fractions, decimals, and percents; and can compute averages. Students can use these skills to solve problems requiring two or more steps and can represent unknown quantities with expressions involving variables. Students can measure length, apply scales, identify geometric figures, calculate areas of rectangles, and are able to use information obtained from charts, graphs, and tables.

$$29 = \square + 16$$

**What number should go in the box to make the number sentence above TRUE?**

ANSWER _______________

**What number does ▼ point to?**

Ⓐ 1

Ⓑ 2

Ⓒ 3

Ⓓ 4

**Here are the ages of five children:**

13, 8, 6, 4, 4

**What is the average age of these children?**

Ⓐ 4

Ⓑ 6

Ⓒ 7

Ⓓ 8

Ⓔ 9

Ⓕ 13

Ⓖ I don't know.

**LEVEL**

**Understand Measurement and Geometry Concepts and Solve More Complex Problems**

**600**

Students at this level know how to multiply fractions and decimals and are able to use a range of procedures to solve more complex problems. Students demonstrate an increased understanding of measurement and geometry concepts. They can measure angles found in simple figures, understand various characteristics of circles and triangles, can find perimeters and areas, and calculate and compare volumes of rectangular solids. Students are also able to recognize and extend number patterns.

**LEVEL**

**Understand and Apply More Advanced Mathematical Concepts**

**700**

Students at this level have the ability to deal with properties of the arithmetic mean and can use data from a complex table to solve problems. They demonstrate and increasing ability to apply school-based skills to out-of-school situations and problems.

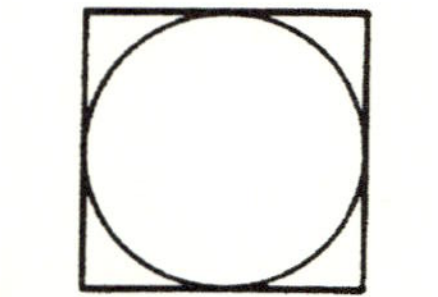

The length of a side of this square is 6. What is the radius of the circle?

Ⓐ 2    Ⓑ 3    Ⓒ 4    Ⓓ 6    Ⓔ 8    Ⓕ 9    Ⓖ I don't know.

**NUTRITIVE VALUE OF CERTAIN FOODS**

|  | Measure | Calories | Protein (grams) | Carbohydrates (grams) |
|---|---|---|---|---|
| Banana, raw | 1 | 100 | 1 | 26 |
| Beef hamburger | 3 oz. | 246 | 21 | 0 |
| Whole milk | 1 cup | 160 | 9 | 12 |
| Doughnut | 1 | 125 | 1 | 16 |
| Eggs, boiled | 2 eggs | 160 | 13 | 1 |

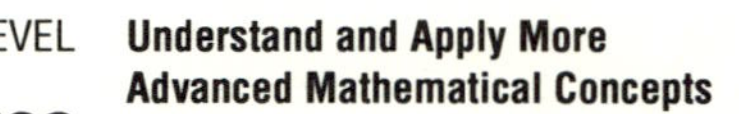

According to the table, what is the total amount of protein contained in two boiled eggs and one-half cup of whole milk?

ANSWER ___________

Figure 3.1.  Proficiency levels and sample items from the 1988 International Assessment of Educational Progress (IAEP). (Lapointe, Mead, & Phillips, 1989. Used with permission.)

mathematics. The official reports from NAEP (e.g., Dossey, Mullis, Lindquist, & Chambers, 1988; Mullis, Dossey, Owen, & Phillips, 1991, 1993) summarized trends in mathematics proficiency using a representative national sample of students, currently from grades 4, 8, and 12 with respect to mathematical processes or abilities (e.g., conceptual understanding, problem solving) and mathematical content (e.g., numbers and operations, geometry) that are common foci in the school mathematics curriculum. In addition to the process- and content-based results, the NAEP reports contained examples of items that appeared on the instrument, information on programs and practices in mathematics instruction, and results from questionnaires on students' attitudes toward mathematics and teachers' mathematics classroom practices.

Like the international assessments, results from the NAEP mathematics assessments are not likely, because of the generality of the survey, to be directly helpful in providing detailed guidance to teachers making instructional decisions in their classrooms. Nevertheless, the general reports of results may provide a fairly good source of information for teachers, mathematics supervisors, and curriculum developers. More extensive reports of NAEP results, including further analyses of student performance and a greater variety of sample items, are often undertaken by NCTM task forces. For example, for the 1986 mathematics assessment, the NCTM published a book, *Results from the Fourth Mathematics Assessment of the National Assessment of Educational Progress* (Lindquist, 1989), and a series of articles written for teachers, which appeared in *Arithmetic Teacher* and *Mathematics Teacher* (e.g., Brown et al., 1988; Kouba et al., 1988).

These latter reports and articles are much more specific about test items and possible interpretations of students' performance than the general NAEP publications. By presenting sample items, together with information about the percent of students who attempted an item and answered it correctly, these secondary analysis reports may provide more helpful information for instructional planning. For example, consider the graphical interpretation and interpolation item shown in figure 3.2. In the official report of the 1986 NAEP mathematics results (Dossey et al., 1988), details regarding student performance on this item are not given, and performance is discussed only generally with respect to a predetermined proficiency level of 300 that involves the ability to "demonstrate more sophisticated numerical reasoning, and . . . to draw from a wider range of mathematical skill areas" (p. 39). In contrast, the secondary analysis of this item (Brown & Silver, 1989) reported

Refer to the following graph. This graph shows how far a typical car travels after the brakes are applied.

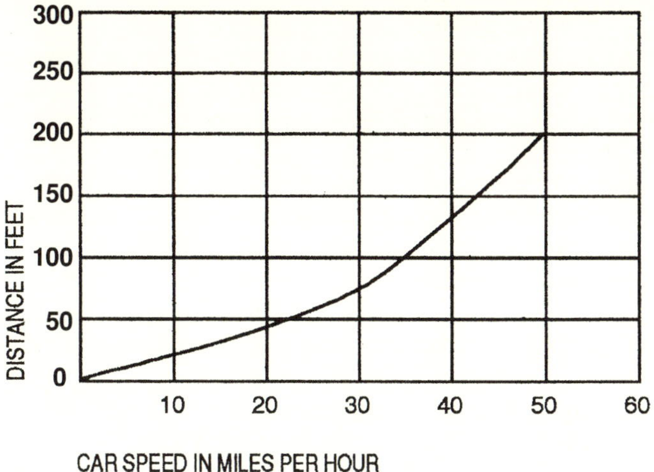


A car is traveling 55 miles per hour. About how far will it travel after applying the brakes?

- ☐ 25 feet
- ☐ 200 feet
- ☐ 240 feet
- ☐ 350 feet
- ☐ I don't know

Figure 3.2. Graphical interpretation item from the Fourth Mathematics Assessment of the National Assessment of Educational Progress (NAEP).

that 41 percent of the grade 7 students and 70 percent of the grade 11 students answered it correctly. Using this kind of information, some teachers might choose to administer this or other "released" NAEP items to their students and then compare their class results to results from the national sample. Additionally, by comparing the percentage of correct figures across grade levels, teachers and other instructional planners and designers can gauge the effects of exposure to the mathematics curriculum and student growth in mathematics with respect to certain content areas and mathematical abilities and identify curricular or process areas that may need greater instructional emphasis.

In the near future, it is likely that the NAEP will continue to be a major source of national assessment information. In fact, recent changes in the NAEP have increased the likelihood that its information will be considered as a source for instructional guidance. By instituting a state-by-state reporting system, and by classifying student achievement both with respect to proficiency scales (Mullis, 1990) and achievement levels (National Assessment Governing Board, 1991), NAEP is moving toward a more visible test and one for which the performance stakes are inching higher. As the country moves toward a national testing system, it is likely that additional sources of national test information will become available.

As national testing becomes more visible and more important, greater pressure will almost certainly be exerted on educators at the state, school district, school building, and classroom levels to use the results as the basis for instructional decisions. With these national data being discussed as sources for instructional decisions, it will be important to examine the relationship between test content and task format and the standards for mathematics curriculum and assessment outlined in the NCTM *Standards* and in other mathematics education reform documents. An analysis of the relationship between the NCTM *Standards* and the content of the grade 8 NAEP mathematics assessment revealed that only about half of the test items were related to the NCTM themes of problem solving, reasoning, and communication (Silver, Kenney, & Salmon-Cox, 1992). Moreover, the analysis also showed that only one item was related to the NCTM *Standards* theme of mathematical connections, that most of the test items were multiple-choice rather than constructed-response tasks, and that the items involving calculator usage were unimaginative. Certainly, one would want to consider these features of the NAEP test and be cautious in making instructional decisions and recommendations based on student performance on this test.

### Summary

Despite the presumption that externally mandated testing should be useful in providing information to improve instruction and learning, experience and research suggests that this is not often the result. In fact, our analysis of these external assessments as sources of information for instructional guidance has indicated the limitations of these tests for providing such information. Whether from international, national, state, or local testing, the results of external assessments offer only limited information on which to base instructional decisions. Because these assessments are far removed from the classroom environment, their results are of minimal utility for detailed instructional guidance, especially for the kinds of interactive decision making (Borko & Shavelson, 1990) that characterize teaching on a daily basis. When suitable at all, the kind of information available from these assessments appears to be best suited for general, long-range planning at the school, district, or state levels. Nevertheless, to the extent that these tests also fail to include broad coverage of rich mathematical content and assess the use of only short-answer and multiple-choice formats, the information will not be very useful to educational professionals who are seeking to reform mathematics instruction.

In the next section, we turn our attention to richer sources of information for interactive decision making and instructional guidance. These sources of information exist in the instructional activities of a mathematics classroom in which students are engaged in the performance of substantive, authentic activities and are reached through use of instructionally embedded assessments that can provide a portrait of students' mathematical proficiencies and competencies—a portrait that becomes evident through observation and evaluation of students as they engage in the performance of the activities.

*Assessment Information Embedded in*
*Instructional Practice: Sources*
*and Resources*

As we have seen, externally mandated tests impinge upon the time available for mathematics classroom instruction, yet the student performance results are quite limited as guides for instructional decisions. In fact, some have argued that the tests are largely superfluous, confusing rather than enhancing teachers' judgments and evaluations. As Hill (1991) has noted: "The teacher is closest to student performance, observes it daily, and assesses it constantly to make instructional decisions. It is doubtful that *any* [external] measure can tell us a fraction of what a teacher ferrets out in the process of instruction" (p. 4).

Earlier in this chapter, we discussed the widespread belief that externally mandated testing has negative effects on classroom instruction. In light of that belief, it is somewhat surprising that surveys of teachers and students have consistently indicated that they believe the educational and psychological effects of classroom evaluation are generally substantially greater than the corresponding effects of standardized testing (Dorr-Bremme & Herman, 1986; Haertel, 1986; Kellaghan, Madaus, & Airasian, 1982; Salmon-Cox, 1981; Stiggins & Bridgeford, 1985). Although externally mandated testing may have a focused, short-term effect on classroom instruction, apparently teachers and students see the cumulative effects of ongoing classroom evaluation as having a greater impact on the learning that does or does not occur and the feelings of satisfaction that do or do not result.

Research has shown that a wide range of evaluative activities takes place in classrooms, with different patterns at different grade levels and in different subject areas (Fennessy, 1982; Gullickson, 1986; Stiggins & Bridgeford, 1985). Activities include evaluation through teacher questioning and class or group discussion, marking or commenting on performances of various kinds, checklists, informal observation of learning activities, written exercises of various kinds (including projects, assignments, worksheets, and text-embedded questions), and teacher-made tests. Although tests and testlike activities constitute only a fairly small component of the total set of evaluation activities in a course, the impact of classroom testing has been studied more extensively than other forms of classroom evaluation. Some studies (Dorr-Bremme & Herman, 1986; Haertel, 1986) have estimated that formal tests occupy about 5 percent of students' time at the elementary school level and about 15 percent at the secondary school level. Mathematics and science teachers, however, have tended to rely more heavily on paper-and-pencil objective tests, whereas teachers in other subjects are reported to rely more on structured observations and professional judgments (Stiggins & Bridgeford, 1985).

Teachers judge evaluative activities to be important aspects of teaching and learning, but they are often concerned about the perceived inadequacy of their efforts (Gullickson, 1986; Stiggins & Bridgeford, 1985). With respect to the evaluation of mathematical problem solving, Silver and Kilpatrick (1988) argued that attempts to control the content and form of teachers' instruction have had the consequence of deskilling teachers by "convincing them that they lack the expertise to assess how their students are learning and thinking" (p. 185).

It is unlikely, however, that a better situation will result from a requirement that teachers make an effort to learn more about the theory of educational measurement. Although it is true that a substantial proportion of teachers, especially elementary school teachers, have little or no formal training in educational measurement techniques, it is equally true that many of those who do have such training find it of little relevance to classroom evaluation activities (Gullickson, 1986; Gullickson & Ellwein, 1985; Haertel, 1986; Stiggins, 1985). Approaches that tie assessment practices to instructional goals and activities in reasonable ways are likely to be more productive. As Silver and Kilpatrick have noted: "What is needed are serious efforts to re-skill teachers, to provide them with not only the tools such as sample problems and scoring procedures that they can use to construct their own assessment instruments but also with the confidence they so often lack in their own ability to determine what and how their students are doing in solving mathematical problems" (p. 185).

The linking of assessment practices to instructional goals seems especially important when we consider the research findings on the content of teacher-made tests and contrast that with the current thinking about the important goals for school mathematics instruction. Most analyses of the content of teacher-made tests have found that the vast preponderance of questions require low-level knowledge and performance on the part of the students. For example, Fleming and Chambers (1983) analyzed 8,800 test questions in twelve grade and subject area combinations (elementary to high school) and found that almost 80 percent of all questions were at the "knowledge" level in Bloom's taxonomy. Even in classrooms in which teachers reported instructional goals involving higher level thinking, Haertel (1986) found that "classroom examinations often failed to reflect teachers' stated instructional objectives, frequently requiring little more than repetition of material presented in the textbook or class, or solution of problems much like those encountered during instruction" (p. 2).

To some extent the limitations in the form of classroom testing may be attributed to the influence of externally mandated tests. Mathematics teachers often create or use multiple-choice and short-answer tests, thereby demanding and evaluating performances from their students only in forms identical to those used on standardized tests. At the elementary school level, especially, many teachers make extensive use of commercially prepared tests that accompany their textbooks. Naturally, these commercially prepared tests neither reflect the instructional nuances of any particular teacher's class nor utilize a rich variety of task formats. Even when textbook tests are not used, teachers often emulate the "tests that really count" and utilize multiple-choice and short-answer formats in their classroom assessment (Fleming & Chambers, 1983; Stiggins & Bridgeford, 1985).

Other reasons have been suggested to account for the poor quality of the content represented in classroom teachers' tests, including the difficulty of writing tasks (especially short-answer or multiple-choice test items) to assess higher level skills (Elton, 1982), the ease with which teachers can defend their grading of students' responses to lower level factual recall items and the resulting higher reliability (Natriello, 1987), and the belief that higher level questions may lead to confusion and frustration on the part of students (Doyle, 1986). Although teachers may have good reasons for orienting their tests primarily toward lower level knowledge and performance, a conflict is likely to exist between these kinds of assessment practices and the objectives of a mathematics curriculum oriented toward higher level thinking, reasoning, and problem solving.

As mathematics classrooms move toward the realization of the vision portrayed by the NCTM *Curriculum and Evaluation Standards for School Mathematics* (1989) and the *Professional Standards for Teaching Mathematics* (1991), they will become environments in which teachers and students work together on making mathematics and on the active exploration of mathematical ideas. As the teaching of mathematics shifts "from an authoritarian model based on 'transmission of knowledge' to a student-centered practice featuring 'stimulation of learning' " (National Research Council, 1989, p. 81), mathematics programs will involve students in a wide variety of activities, such as

- working collaboratively;
- asking and answering questions posed by fellow students or the teacher;
- engaging in substantial discussions about mathematics;
- thinking hard about what they are learning and about the nature of mathematics;
- working on extended projects that may take days, or even weeks, to complete;
- solving worthwhile and challenging problems on teacher-made mathematics tests and on homework assignments.

These and other activities have the potential to function as instructionally embedded sources of assessment information that can be used for instructional guidance as well as for summative evaluation of students' achievement. The next section of the chapter discusses suggested sources of instructionally embedded assessment opportunities within the venues of classroom discourse and activities and the direct performance of mathematical tasks, as well as the kinds of information that these classroom assessments can provide for teachers and other decision makers.

## ASSESSMENT EMBEDDED IN CLASSROOM DISCOURSE AND ACTIVITY

The NCTM *Professional Standards for Teaching Mathematics* (1991) places a heavy emphasis on the role of discourse in facilitating students' learning of mathematical ideas: "The discourse of a classroom—the ways of representing, thinking, talking, agreeing and disagreeing—is central to what students learn about mathematics as a domain of human inquiry. . . . Students must talk, with one another as well as in response to the teacher. When the teacher talks most, the flow of ideas and knowledge is primarily from teacher to student. When students make public conjectures and reason with others about mathematics, ideas and knowledge are developed collaboratively, revealing mathematics as constructed by human beings within an intellectual community. Writing is another component of the discourse" (p. 34). In classrooms that are moving toward an embodiment of the NCTM *Standards*, the classroom discourse—between teacher and student and among students—is centered around worthwhile mathematical tasks, and the intellectual activity of the students provides a rich environment from which assessment information can be obtained. A teacher can gain valuable information for instructional guidance by watching students as they work on mathematical tasks, by observing students working in pairs or in groups, by asking appropriate questions at opportune moments, and by listening to students present their answers or solutions, their approaches or methods, and their explanations or justifications.

Although the activity and the discourse can certainly serve as a rich source for teachers' interactive instructional decisions, because discourse and intellectual activity in the classroom are ephemeral entities, much of the assessment information that could be gained is likely to go unrecorded and therefore remain unavailable for instructional planning and long-range decision making. In this section we give some attention not only to a few forms in which assessment information might be extracted from naturally occurring classroom discourse and intellectual activity but also to some ways of preserving records of the discourse and student activity in order to facilitate instructional planning.

*Observation.* Watching students while they are "doing mathematics" can provide insights into their understandings and misunderstandings. Observation is almost certainly the most basic classroom process of gathering assessment information about students. Information gained from careful observation is regularly used by teachers to decide whether to move forward in a lesson or give more time for completion of a component activity, to decide whether to provide an additional example or a different explanation, and to modify the expected direction of a lesson or a unit of instruction.

Although the number of students that can be observed at any one time is limited, observation is an assessment method that is generally comfortable and convenient for classroom teachers because it is relatively easy to include as part of regular classroom routines, and it is useful for assessing a range of student characteristics, including performance, attitudes, and beliefs. Although record keeping can be cumbersome without advanced planning, recording schemes for systematic observation are relatively easy to construct, and many sources of ideas for observation instruments for the mathematics classroom exist (e.g., British Columbia Ministry of Education, 1990; Charles, Lester, & O'Daffer, 1987; Stenmark, 1989).

Among the many types of observation instruments that exist, two types are particularly appropriate for use by the classroom teacher: the annotated class list and the topical list. These instruments are easy both to construct and administer, but their purposes differ somewhat. The annotated class list consists of a roster of student names with a space to the right of each name for recording a variety of student attributes such as mathematical understanding (or misunderstanding), demonstrated attitudes toward mathematics, and potential areas in which the student excels or needs assistance. The topical list consists of a set of predetermined categories to be used during the observation. For example, during an opportunity to observe students solving problems, a teacher might choose these categories as the focus for her observation of students:

- tries to understand the problem,
- selects appropriate solution strategies,
- shows a willingness to switch between solution strategies,
- uses a systematic procedure,
- shows perseverance,
- checks work and answer.

Teachers can use information from documented classroom observations in a variety of ways to assist in instructional decision making. In an unobtrusive way teachers can watch and listen to students as they explain their mathematical thinking and work in groups, thus gaining a feel for the students' facility with communication. In some cases a student might demonstrate more understanding than that indicated on a written test or on homework. Teachers can get a sense of how students have processed and interpreted information about the area of mathematics under consideration—or how they have not processed or misinterpreted the information presented. Taking time to observe students at work, then, can prove opportunities for timely feedback that can shape decisions to be made regarding instruction in the next class period and beyond.

*Questioning.* It is difficult to imagine a mathematics classroom without questioning as a central activity, and it is difficult to imagine instructional decisions that are not informed in some way by students' responses to questions. Questioning—both planned and unplanned—can be a source of useful information for student assessment related to students' cognitive performance, attitudes, beliefs, mathematical insights, and metacognitive processes; and students' responses to questions can provide a valuable source of information for interactive instructional decisions involving adjustment of lesson pacing, example selection, homework assignments, and so on, as well as for longer term instructional planning decisions.

The idea of gaining assessment information through questioning students is far from novel. An extensive literature exists on classroom teacher questioning practices. Carlsen (1991) presents an excellent review of these studies with respect to the context and content of teacher questions and the responses of teachers and students to questions. Much of this work has been done from the perspective of the process-product paradigm (Mitzel, 1960; Rosenshine & Furst, 1973), in which student outcomes (usually, though not exclusively, student achievement) are viewed as a function of discrete, observable teacher behaviors, or from the sociolinguistic perspective (Cazden, 1986; Green, 1983), in which teacher questions are viewed as a mutual construction of teachers and students, rather than being exclusively the result of teacher generation; and the research focus has been on the linguistic character of the communication structures and the social dimensions of the questioning, such as its role in reflecting and reinforcing authority and social status relationships in the classroom.

Although they appear to be designed to enhance student discourse, teacher questions may sometimes discourage students from speaking. For example, Dillon's (1985) analysis of five classrooms showed that teacher questions typically produced terse, factual statements by students, whereas noninterrogative expressions produced lengthier, more syntactically complex responses. He concluded that teacher questions in these classrooms had the consequence of suppressing rather than enhancing student discussion.

Given the new vision for mathematics instruction in which teachers are to pose questions that "elicit, engage, and challenge each student's thinking" (NCTM, 1991, p. 35), discourse-inhibiting questioning such as the kind described by Dillon (1985) is unacceptable. Recent publications (e.g., Bennett & Foreman, 1990; Stenmark, 1989) have included sections on teacher questioning in the mathematics classroom and sample questions in the areas of problem comprehension (What is this problem about?), relationships (Is there a pattern?), communication (Could you explain what you think you know about this concept right now?), and self-assessment (What kind of mathematics problems are still difficult for you?). Questions such as these would appear to have promise in enhancing student-teacher discourse, thereby providing teachers with important information for instructional guidance.

Although direct questioning of students can certainly be a source of information for instructional decision making, its level of importance may change as the mathematics classroom environment evolves to meet the NCTM *Curriculum and Evaluation Standards* and the *Professional Standards for Teaching Mathematics*. As teachers become facilitators rather than interrogators and as teachers begin to use other kinds of mathematics tasks to elicit students' higher order thinking skills, the use of direct questioning of students at the classroom level will likely diminish.

One practice in which teachers' use of direct questioning is likely to remain important, however, is that of individual student interviews. Structured or semi-structured interviews, in which a preselected problem situation and a set of probing questions are used, have long been used by researchers to study students' mathematical performance and the extent of their understanding of mathematical concepts and procedures (e.g., Erlwanger, 1973). Classroom teachers can also use this form of assessment to collect and record detailed information about students' mathematical understanding and problem-solving processes. As Peck, Jencks, and Connell (1989) have suggested, "Just as student interviews have been helpful in uncovering conceptual difficulties, they can be a useful tool for guiding the progress and direction of day-to-day classroom work" (p. 15).

The key to a successful interview assessment is a well-designed interview plan. Although such plans may vary depending upon the problem situation presented, they usually are composed of six steps: establishing rapport, presenting specific instructions, presenting the problem, probing for understanding of the problem, probing for the solution process, and coming to closure (British Columbia Ministry of Education, 1990; Charles, Lester, & O'Daffer, 1987). During the course of the interview, two principles of importance are acknowledged: sufficient time should be allowed for the student to formulate a response, and the student's thought processes should be of greater importance than the answer.

Whether the questioning situation occurs informally as part of regular classroom activity or in a more structured individual or small group setting, teachers can gain insights into students' thinking and communication skills that may not be obviously apparent from written work. Questioning and interviews, as forms of oral discourse, can also provide diagnostic information so that teachers may direct instruction to review concepts that prove to be problematic or to review briefly those areas in which students have demonstrated understanding.

*Written Discourse.* As the earlier quote from the NCTM *Professional Standards for Teaching Mathematics* indicated, not all discourse is oral; various forms of written discourse can also serve well as sources of information for instructional guidance. For example, student journals, or students' written responses to specific probes concerning their learning or their disposition, provide opportunities to consider students' ideas or their attitudes and beliefs, when making instructional planning decisions.

Rose (1989) suggests that journal writing in mathematics classrooms can be instrumental in setting up a useful dialogue between students and their teachers: "Students and teachers find something to talk about and the classroom becomes more cooperative and humanized as each see [sic] the other in a new and personalized light" (p. 25). The function, content, and format of the journal depends upon its intended use as part of classroom assessment. For some purposes, journals might be used as a repository for students writing in an expressive mode, in which students "think aloud on paper" and record their impressions of mathematics classroom activity or their learning. In other cases, students might be asked to use their journals for transactional writing, in which a journal entry serves as a record of specific responses to a teacher's questions or provides some designated kinds of information. In either case, the journal serves not only as a record of the students' thoughts but also as a medium for dialogue with the teacher. A powerful dialogue between student and teacher can occur during the typical journal-writing sequence: student entries followed by the teacher reading, reflecting, and commenting on the entries, followed by new student entries, and so on. As teachers use the journals to understand the thinking, feelings, or recorded observations of their students, they can apply this information when making instructional decisions.

For a variety of reasons, but certainly because journals are perceived both to use valuable classroom instructional time and to require additional reading and commentary from teachers, relatively few mathematics teachers use journal writing as a major instructional activity in their classrooms. Nevertheless, some form of written discourse is possible in all mathematics classrooms, and the frequency of occurrence is likely to increase in the next decade. As an alternative to full-scale journal writing, teachers might regularly have students respond in writing to specific probing questions, or to complete "sentence starters" (Stenmark, 1989) such as "Today in mathematics I learned . . . " or "Of the math we've done lately, I'm most confused about . . . " Clearly, students' responses could serve as a valuable source of information to guide planning for the following lesson or week's work, and if students were asked to respond to such probes on a regular basis, this information would in all likelihood become an important component of instructional decision making and could provide teachers with important information about how students are thinking and feeling about their mathematics classroom experiences.

Another example of a writing task that provides assessment information for instructional guidance is one that might be directed toward a specific mathematical concept. In *Thinking Through Mathematics*, Silver, Kilpatrick, and Schlesinger (1990) present an account of a teacher who, before beginning a unit on geometry, used a simple statement ("Tell me everything you know about circles") as a means to find out how much her tenth graders remembered about this geometric entity. Students' knowledge was found to vary from remembering simple facts ("A circle is round"), to remembering formulas ($A = \pi r^2$), to more sophisticated ideas ("It's made up of a series of arcs that are all connected"). This simple exercise provided the teacher with information that allowed her to structure the lesson on circles, to identify students who might need additional help on this topic, and when she asked the same question again at the end of the unit, to identify how the students' understanding of circles had changed as a result of instruction. Thus, by giving students a few minutes to respond to a simple question or statement at an opportune time, teachers may gain important information that can enhance their instructional decision making.

Research has shown that teachers consider both cognitive and noncognitive information in making their instructional planning decisions (e.g., Shavelson & Stern, 1981). Through observation, conversation, and reading students' journals and other writings, teachers may be able indirectly to gain valuable information about students' dispositions toward mathematics and toward themselves as learners of mathematics. This information can also be gained more directly through the use of attitude or belief surveys, examples of which appear in many recent publications (e.g., British Columbia Ministry of Education, 1990; Charles, Lester, & O'Daffer, 1987; Mullis et al., 1991; Nicholls, Cobb, Yackel, Wood, & Wheatley, 1990; Stenmark, 1989). By having students write responses to a set of statements (e.g., "In mathematics, memorizing is more important than creative thinking"; "I will keep working on a problem until I get a right answer") in a simple "yes or no" format or with a more complex scale (Strongly Disagree, Disagree, Neutral, Agree, Strongly Agree), or by having students respond directly to questions or statements (e.g., "What is the biggest worry affecting your work in mathematics class at the moment?" "In mathematics class I like . . . because . . . "), teachers can view the development of students' thinking and attitudes from the students' perspective. Moreover, responding to attitude and belief surveys can foster in students a tendency to engage in self-assessment, a proactive, internal process that promotes the development of mathematical power (Kenney & Silver, 1993).

### Information from Direct Assessment of Performance on Mathematical Tasks

Although the prior section has dealt with observation and discourse that occurs around and about mathematical tasks being used in the classroom, the discussion thus far has focused more on the nature of the discourse itself and less on the tasks and the criteria for judging performance. In this section, we consider the instructional guidance information provided by various types of mathematical classroom tasks and the associated judgment criteria.

*Projects and Open-Ended Problems.* In contrast to the short-answer or multiple-choice questions that typically make up mathematics achievement tests, opportunities for students to engage in extended exploration of mathematical ideas and situations are provided by projects, investigations, and open-ended problems. As such, they also provide opportunities for teachers to assess their students' abilities to formulate problems, to apply their knowledge in novel ways, to generate interesting solution approaches, and to sustain intellectual activity for an extended period of time. Moreover, by working on an extended investigation of an interesting mathematical problem, students participate in activities that are closely related to the nature of complex performances outside of school, through which they can develop an understanding that the analysis of complex problems may take days or even weeks to explore; they can learn to work independently or collaboratively on a large project; and they can experience the process of producing a written or oral report of their work over an extended period of time. Open exploratory tasks have been frequently used in mathematics instruction in other countries—for example, Australia (Clarke, 1988) and Great Britain—but they are infrequently used in the United States (Dossey et al., 1988; Mullis et al., 1991). As ideas central to mathematics education reform become more prevalent in mathematics teaching, one would expect their use to increase.

An example drawn from one use of mathematical investigations in Australia may illustrate the value of this kind of task in mathematics teaching. Stephens and Money (1993) reported the use of investigation and open-ended problems by the Victoria Curriculum and Assessment Board as part of its external examination of students for high school credit in their mathematics courses. One component of the examination is an investigative project representing fifteen to twenty hours of student work, and another component is a challenging problem chosen from a set of four problems and requiring about six to eight hours of student work. For our purposes here, what is interesting about this example is the fact that such problems are viewed by this external examination board as providing important information about students' mathematical attainments and that these novel tasks are administered by classroom teachers, who are required to allow students to work on the investigative project for seven to ten class periods and who then score the students' work using grading criteria organized around three aspects of the work:

- Problem definition (clear definition of requirements, assumptions, variables, and identification of the nature of the solution being sought);
- Solution and justification (production of a solution, appropriate use of mathematical language; accuracy, interpretation of results; depth of analysis; quality of justification of solution);
- Solution process (relevance of mathematics used; generation and analysis of appropriate information; recognition of relevance of embedded findings, refinement of problem definition).

The list of criteria suggests the kinds of information that teachers could obtain from the use of such tasks as part of mathematics instruction. In fact, according to Stephens and Money (1993), the teachers involved with the external examination program use the assessment criteria and the related grade descriptors as a basis for planning their teaching program. One can easily imagine that teachers in the United States could use similar criteria to guide instruction, even if the tasks were not part of an external examination program but rather were generated and implemented by mathematics teachers.

Especially, though not exclusively, for younger students, it may be desirable to have the results of projects and open-ended problem investigations reported orally as well as, or rather than, in writing. The combination of oral and written presentations would allow for a more thorough evaluation of students' mathematical performances, and similar criteria could be used to evaluate oral and written presentations. In this way, even elementary school teachers would be able to obtain assessment information about students' mathematical thinking that could help guide instructional decisions.

The publication of a number of sources of open-ended problems and projects (e.g., Shell Centre for Mathematics Education, 1984; Souviney, Britt, Gargiulo, & Hughes, 1990; Trowell, 1990), and some published discussions of the pedagogical rationale for activities such as open-ended problems (e.g., Silver & Adams, 1987; Silver, Kilpatrick, & Schlesinger, 1990; Silver & Mamona, 1990) provide mathematics teachers with a base from which to include these kinds of activities in instruction. Once included as a regular feature of mathematics teaching, the activities should become a valuable asset in obtaining student performance information to guide instruction.

*Classroom Testing.* Written tests are commonly used in classrooms to assess each individual student's achievement in mathematics. As noted earlier, research has shown that formal classroom testing occupies a substantial portion of instructional time (5 to 15 percent). In addition to their role in providing summative information on student achievement, classroom tests also represent a major source of formative feedback that might be useful in guiding instructional decisions. Unfortunately, their value as instructional guides is limited if, as has been indicated in the research discussed earlier, the tests constructed by mathematics teachers tend to make heavy use of short-answer and multiple-choice formats. Unfortunately, an excessive emphasis on short-answer questions creates the dual impression that only the final answer matters and that what is valued in mathematics is the ability to answer many questions quickly, almost certainly as a result of having memorized many facts and procedures that can be recalled rapidly and applied flawlessly. However, to the extent that teachers and other instructional decision makers come to view mathematical activity in a manner consistent with reform documents like the NCTM *Curriculum Standards* and the *Professional Standards for Teaching Mathematics*, as a process primarily involving reflective reasoning, problem solving, and communication about mathematical ideas, it is clear that classroom testing will need to include formats other than questions requiring only short answers or the choice of one answer from a set of options.

Teachers interested in diversifying their classroom testing might include a project or an open-ended problem, such as that discussed previously, as a "take-home" portion of a test. The inclusion of such an activity as a portion of a test would provide a teacher with information about aspects of student performance that could not be made available solely from classroom testing. Even within the time constraints characteristic of classroom testing, however, it is possible to include tasks that can provide information on students' reasoning, problem solving, and communication. This can be accomplished through the use of mathematical tasks that can be completed in 5 to 15 minutes, in contrast to the 30 to 45 seconds typically available for a response to a multiple-choice question, and that bear a "family resemblance" to projects and extended open-ended problems that require much longer to complete. Such tasks are being used in external testing programs in mathematics such as the College Board Advanced Placement (AP) Test in calculus and in some state-level testing programs (e.g., California, Connecticut, Kansas, Maine). The California Assessment Program (CAP) has used these kinds of problems in its mathematics assessment for several years; it has published examples of the tasks, the scoring guides, and sample student responses in *A Question of Thinking* (California State Department of Education, 1989) and *A Sampler of Mathematics Assessment* (Pandey, 1991).

An example of this kind of task drawn from the assessment developed for the QUASAR project[4] may illustrate its value as a source of information for instructional guidance. A sample QUASAR assessment task appears in figure 3.3. A sample answer to this task might be "No" with an explanation such as "Yvonne takes the bus eight times a week, and this would cost $8.00. Because the bus pass costs $9.00, she should not buy the pass." It is possible, however, that a student might answer "Yes" and provide a logical reason, such as "Yvonne should buy the bus pass because she rides the bus eight times for work and this costs $8.00. If she rides the bus on weekends to go shopping, it would cost $2.00 or more, and this would be more than $9.00, so she can save money with the bus pass."

When these tasks are used as part of external testing programs, student responses are typically scored holistically, using a scoring guide that provides detailed information about various levels of performance in solving a particular problem. In QUASAR, for example, students' responses are scored using a scoring guide (rubric) that attends to three categories of solution characteristics (Silver & Lane, 1993):

- Mathematical knowledge (knowledge of relevant concepts, procedures and principles; identifying relationships among

The table below shows the cost for different bus fares


BUSY BUS COMPANY FARES


One Way                              $1.00
Weekly Pass                          $9.00

---

Yvonne is trying to decide whether she should buy a weekly bus pass.
On Monday, Wednesday, and Friday, she rides the bus to and from work.
On Tuesday and Thursday, she rides the bus to work but gets a ride home with
her friends.

Should Yvonne buy a weekly bus pass?

Explain your answer:

Figure 3.3. Sample QUASAR task.

problem elements; identifying and executing appropriate procedures; verifying results; integration of mathematical ideas),

- Strategic knowledge (appropriate use of mathematical models, including diagrams and symbols; use of appropriate problem-solving strategies; systematic application of strategies),
- Communication (appropriate expression of mathematical ideas in words, mathematical symbols, or pictorial representations; reasonable use of vocabulary, mathematical notation and structure to represent ideas; quality of justification of a solution).

These categories of evaluation criteria illustrate the kind of information that can be obtained both for the summative evaluation of students' achievement and for the purposes of instructional guidance.

Tasks from programs such as CAP and QUASAR can provide examples that mathematics teachers can use or modify for other grade levels. Teachers who use open-ended tasks as part of formal assessment can derive a wealth of information from student responses. As we have seen, they are able to learn whether or not their students can recognize the main points of a problem, organize information, interpret results, use appropriate mathematical language, and express their own thinking and reasoning processes (Stenmark, 1989). However, reading lengthy responses to these tasks is far different from checking multiple-choice responses or scoring the more typical computational responses to mathematics problems. Moreover, the development of detailed scoring rubrics that set forth requirements for varying levels of performance may be essential to ensure high degrees of interrater agreement when the tasks are used on external assessments; however, classroom teachers are unlikely to have the time to create such detailed scoring guides. For classroom use, scoring rubrics need not be as elaborate or detailed as those used in external testing programs, but they can still provide teachers with a mechanism for evaluating students' solutions and examining the evidence provided in the students' response to detect clues that might help guide instructional decision making. For example, a teacher might focus on strategy selection and use, and then use that information to plan additional instruction or to choose examples for the next unit. If such tasks were employed on a regular basis, the accumulated information about students' strategy selection and use, for example, might suggest curricular or instructional changes that could be implemented in subsequent years.

*Homework and Other Assignments.* In addition to their use on classroom tests, the kinds of tasks discussed in the previous section can serve as excellent in-class or homework assignments. In whatever capacity they are used, they can provide information that teachers can use to plan instruction. To the extent that homework and in-class assignments can be used to provide students with a wide variety of mathematical experiences—experiences that emphasize mathematical problem solving, reasoning, and communication—the information gathered by teachers through an examination of students' performance can enrich instructional decision making. If the homework and in-class assignments are restricted to "blue ditto sheets" or "more-of-the-same" exercises to be solved using well-rehearsed procedures, the information obtained will be of very limited value for guiding instruction in the direction suggested by documents like the NCTM *Professional Standards for Teaching Mathematics.*

### Accumulating Instructional Guidance Information

In our discussion we have noted several times that much of the information obtained from instructionally embedded assessment can be especially useful for instructional guidance if it is accumulated over time. Careful record keeping can help to ensure that longitudinal information is accumulated for examination. A technique that has been suggested as particularly appropriate for accumulating instructionally embedded assessment information is the mathematics portfolio. In its most general sense, a portfolio is "a container of evidence of someone's knowledge, skills and dispositions" (Collins, 1990, p. 159). Creation of a collection of work produced over time has long been an accepted form of assessment in the arts and humanities, and it has received some attention in mathematics in recent years (Mumme, 1990). In general, attention to portfolios has focused on the use of this technique as an alternative or supplement to formal testing as a means for evaluating student achievement. For our purposes here, it is more important to focus on portfolios as a source of useful information for instructional guidance.

Although portfolios could conceivably be assembled for a variety of purposes, the two purposes most often discussed are to display the "best work" or demonstrate "growth over time." The purpose of the portfolio determines the criteria that will be used in selecting its contents. If the portfolio is meant as a summative display of proficiency, then the samples representing a student's best work are most appropriate for inclusion. In contrast, if the purpose is for documentation of growth and progress over time, then it would be desirable for it to contain dated examples of student work, including drafts and final copies of projects, solution attempts (both successful and unsuccessful) for a particular problem, and perhaps some examples of the use of concepts or procedures early in a course as contrasted with their use later in the course.

Most discussions of classroom use of portfolios emphasize the benefits of involving students actively in the selection of items to include in their portfolios. In this way, they can engage in an important process of self-assessment and the portfolio can be further personalized, such as when a student who is interested in art decides to include examples of pictorial representations of and solutions for mathematics problems. The following (nonexhaustive) list compiled by Stenmark (1989) suggests the wide range of contents that might be included in a mathematics portfolio:

- Written descriptions of the results of practical or mathematical investigations, pictures and dictated reports from younger students;
- extended analyses of problem situations and investigations;
- descriptions and diagrams of problem-solving processes;
- statistical studies and graphic representations;
- reports of investigations of major mathematical ideas;
- responses to open-ended questions or homework problems;
- group reports and photographs of student projects;
- video, audio, and computer-generated examples of student work.

For younger children, or for students who are compiling a portfolio for the first time, the teacher might maintain the portfolio for each student and periodically review it with the student until the student becomes more familiar with the process (Collins, 1990).

The states of California and Vermont have taken an active role in the development of portfolio assessment and have each published examples of student work as well as scoring guidelines. The California Assessment Program investigated the use of two categories for scoring individual portfolios—evidence of mathematical thinking (e.g., organization of data, conjecturing, exploring) and the quality of activities and investigations (e.g., evidence of significant investigations, connections between mathematical content areas)—and one category for evaluating the portfolios from an entire class based on the variety of approaches and investigations used across the set of portfolios (Mumme, 1990; Pandey, 1991). In its report of results from the 1990–1991 pilot project, the state of Vermont (Vermont Department of Education, 1991) published the scoring guide for the "best pieces" component of the portfolio project along with sample student responses at some of the score levels. The scoring criteria were based on two elements—problem solving and mathematical communication—and on these criteria within each element:

- Problem solving (understanding the task, quality of approaches and procedures, decisions along the way, outcomes of activities);
- Mathematical communication (language of mathematics, mathematical representations, clarity of presentation).

For the classroom mathematics teacher the portfolio provides a comprehensive view of students' mathematical experiences over time that combines the advantages of other forms of instructionally embedded assessment. Virtually all of the classroom assessment information sources that have been discussed in this section—student journals, written responses to open-ended problems, summaries of group projects or independent investigations, homework papers—can be included in a portfolio. The coupling of portfolios with information from classroom discourse and activity (e.g., observations and interviews) constitutes a multifaceted approach to evaluation of individual student achievement and a vast reservoir of information for making instructional decisions.

Although information from portfolios and other sources can be directly beneficial to classroom teachers, it is less clear how this information can be transmitted beyond the classroom door and into the hands of other education professionals who also engage in instructional decision making. However, there are some ways in which other teachers, curriculum developers, and mathematics specialists can benefit from instructionally embedded assessment information. For example, an eighth-grade mathematics teacher could receive the portfolios that her students compiled last year in their seventh-grade mathematics classes. By looking over students' prior work in mathematics, the new teacher could make preliminary diagnostic decisions regarding strengths and weaknesses on previous mathematics content and, perhaps, make some preliminary predictions regarding areas in which students may experience success or difficulty. Student portfolios might also be useful to curriculum developers and mathematics supervisors. For example, a curriculum development committee could devote a series of meetings to studying a sample of student portfolios at a particular grade level to determine the extent of content coverage. Mathematics supervisors could examine a set of portfolios on the basis of the diversity of activities included, and they could plan staff development activities directed at potentially interesting sources that might not have been included (e.g., projects, journals).

## TOWARD A VISION OF INSTRUCTIONALLY GUIDED ASSESSMENT AND ASSESSMENT-GUIDED INSTRUCTION

In this chapter we have tried to view assessment, whether originating outside or inside the classroom, as an important source of information for instructional decision making in mathematics. In our review, we have pointed out the generally limited value of externally mandated tests in providing instructional guidance information for classroom teachers, but we have also noted some ways in which the information from such testing might be useful at a more global level to other instructional decision makers, such as administrators, curriculum developers, or supervisors. On the other hand, we have seen that a rich instructional program in mathematics has embedded within it a great deal of useful assessment opportunities and information for classroom teachers, yet it is more difficult to see how this instructionally embedded assessment information could be shared with others outside the classroom. A fundamental challenge, then, is to blend the best of both types of assessment to develop approaches that allow the display of a wide array of assessment information that can be helpful both to classroom teachers and other instructional decision makers.

In our consideration of external assessments, we noted that these tests often have little to offer in the way of instructional guidance and that when they affect instructional practice the effects are often judged to be negative. Unfortunately, this theme is not a new one. More than a decade ago, a National Institute of Education (1979) conference report on testing and instruction contained the following statements: "Current testing procedures are not helpful to teachers or students in their day-to-day efforts to teach and learn" (p. v) and "Present day testing programs are largely extraneous to everyday classroom teaching" (p. 359). Given the interest in improving the relationship between assessment and instruction, there have been many calls for changes in testing practice that would result in test formats being aligned more with instructional tasks and in test results that would be more useful for instructional decision making (e.g., Glaser, 1986; Linn, 1983; Nitko, 1989). Some have noted that progress in the field of cognitive psychology offers a new perspective on educational measurement and evaluation with respect to innovative assessment design models (Snow & Lohman, 1989) and the characterization and measurement of skilled performance (Glaser, 1986, 1988). The kind of assessment that is envisioned would involve both assessment-guided instruction and instructionally guided assessment.

To date, the most prevalent approach to aligning assessment efforts with instruction in mathematics has been to focus on assessment-guided instruction by altering the content and form of externally mandated tests. In particular, noting the reported tendency of teachers to be influenced by the content and form of externally mandated tests (e.g., Madaus et al., 1992), many educational reformers have advocated an assessment-driven reform strategy. The major premise of this strategy can be called the *what you test is what you get* (WYTIWYG) principle; that is, teachers will devote substantial instructional attention to the subject-matter content and item formats represented on externally mandated tests. Therefore, some educational reformers have concentrated on substantially altering these tests, with the hope of thereby influencing classroom instruction to move in desirable directions. The results of these efforts are now appearing in some state-level and district-level testing programs. For example, as noted earlier in this chapter, some states have developed mathematics assessments that not only attempt to measure a range of curricular topics broader than that covered in the typical classroom in the state, but they also utilize non-multiple-choice formats, including open-ended tasks (e.g., California), collaborative group assessments (e.g., Connecticut), and portfolios (e.g., Vermont). The content and performance goals of these tests are compatible with documents like the NCTM *Curriculum Standards*, and the tasks and activities used in these assessments are not unlike those mentioned earlier in this chapter as characteristic of a strong classroom mathematics instructional program. There are at least two expected benefits of influencing teachers to "teach to the test" in this case: teachers may begin to use the alternative measures in their own assessments, and the test results may be more useful to the teachers for instructional guidance.

There are, however, some limitations to an assessment-driven reform strategy such as WYTIWYG. Silver (1992) argues that the "teaching to the test" phenomenon may not be sufficiently robust to support the reform effort in mathematics assessment. He further argues that a more realistic view would be "what you get is what I can teach" (WYGIWICT). Teachers, especially elementary school teachers with limited knowledge of and experience with mathematics generally tend to feel more comfortable with and capable of teaching lower level knowledge and skills rather than more complex knowledge and processes. These teachers are quite likely to be influenced less by higher level content on tests than the assessment-driven reform advocates might hope. Because much of the evidence pointing to the influence of tests on instructional practice has found the influence to be in the direction of basic skills, which is the direction also predicted by WYGIWICT, it is difficult to predict the impact that higher level testing alone could have on mathematics teaching. Moreover, some reports on teachers' beliefs and actions (e.g., LeMahieu & Leinhardt, 1985; Salmon-Cox, 1981) contain evidence that the relationship between testing and teaching is more complex than that implied by WYTIWYG, because teachers are not always greatly influenced by the content or format of standardized or externally mandated achievement tests. Given this view, it is unlikely that an assessment-driven reform strategy can be successful without attention also to the continuing education and support of teachers to fortify their classroom instructional programs. Any approach that emphasizes external assessments and ignores the instructional programs is unlikely to succeed in making substantive improvements in the teaching and learning of mathematics.

The vision of merging assessment-guided instruction and instructionally guided assessment will be realized if classroom mathematics instructional programs are strong, if attention is given to making good use of the naturally occurring opportunities to collect assessment information that are embedded in instruction, and if externally mandated assessments become instructionally guided, that is, closely tied to important instructional goals. The likelihood of this merger occurring will be enhanced if viable models and approaches can be identified and implemented.

The example from Australia discussed earlier in this chapter may be a good model to consider because of the approach it takes to blending externally mandated assessment with classroom instructional needs. According to Stephens and Money (1993), the Victoria Curriculum and Assessment Board requires an external examination of students for high school credit in their mathematics courses. The examination consist of four parts: (1) an investigative project representing 15 to 20 hours of student work, (2) a challenging problem chosen from a set of four problems and requiring about 6 to 8 hours of student work, (3) a 90-minute, multiple-choice, midterm test of skills and standard applications, and (4) a 90-minute final test requiring solution of four to six problems, some of which are routine and others nonroutine. Thus, this examination demands several different types of student performance and assesses a wide range of mathematical proficiencies associated with a course of study. As noted earlier, some parts of these assessments are administered and scored by classroom teachers, and teachers are required to devote some portion of their classroom time to parts (1) and (2), which require a substantial time commitment on the part of students. Because the scoring criteria for these challenging tasks are shared with teachers and because the tasks themselves become part of the instruction during the course, the external exam becomes a focus of instructional attention. Yet, because the tasks represent important and desirable educational outcomes, the instructional influence is appropriate. Moreover, the examination design is heavily influenced by a consideration of important instructional goals.

Another model approach to consider is the use of portfolios. As they have been used in Vermont's pilot project, portfolios represent another way to blend the needs of externally mandated assessment with classroom instructional needs. Like the Australian examination system, the assessment plan under development in Vermont consists of several different kinds of student performances: "best pieces" of student work identified by a student, a broader compilation of a mathematical work, and a uniform test of a student's knowledge and understanding of mathematics concepts and procedures (Vermont Department of Education, 1991). Classroom teachers were heavily involved in the Vermont portfolio pilot study in a variety of ways, such as providing the tasks and activities that were included in their students' portfolios and helping their students select the "best piece" examples. Teachers also participated at the state level by serving on the committee that set the portfolio scoring criteria and by participating in the rating sessions for the sample portfolios. It should be noted, however, that a number of technical problems have been detected in Vermont's use of portfolio assessment to report student performance (Koretz, Klein, McCaffrey, & Stecher, 1993) and these technical problems must be solved before the portfolio model is formally adopted in Vermont. Nevertheless, despite the technical problems encountered in Vermont's externally mandated, state-level assessment program, it appears that the portfolio model has considerable potential for merging the interests of assessment and instruction. In particular, as portfolio evaluation guidelines are made available to teachers and as teachers are trained to evaluate portfolios, it is likely that teachers will devote instructional attention to the variety of mathematical activities that are embodied in the portfolio process, thereby promoting instructional reforms at the classroom level.

Neither the Victoria examination nor the Vermont portfolio approach is likely to be viewed by all as viable and desirable. Yet, the need to develop alternative approaches to assessment that can serve the needs of external accountability and internal instructional guidance are absolutely critical. If the efforts of this wave of mathematics education reform are to succeed, assessment and instruction will need to become meshed to a greater extent than ever before. Tests will need to become viewed as providing vital information for instructional guidance. In this way, we may see the realization of Glaser's futuristic prediction: "In the twenty-first century, tests and other forms of assessment will be valued for their ability to facilitate constructive adaptations of educational programs" (1986, p. 45). Moreover, as instructional programs are enriched to engage students actively in the consideration of mathematical ideas, with an emphasis on problem solving, reasoning, and communication, teachers will need to be more skilled in extracting important information from the on-going activities in the assessment-rich environment of the classroom.

As the nation rises to the challenge of establishing national standards and national testing programs to monitor students' progress toward those standards, it is more important than ever to remember that the kinds of tests we need are those that can also serve well as guides for instruction. It would also be wise for us to recall that instructional activities in the classrooms of good teachers can be a rich source of assessment information—richer by far than any single test both for measuring student achievement and also for informing and guiding instructional decisions. Our educational system is unlikely to be improved by designing and implementing a national examination system unless that system has at its heart solid instructional goals for students and a sensible approach to assessment that blends internal and external forms of assessment information and that includes attention both to instructionally guided assessment and to assessment-guided instruction.


NOTES


1. The preparation of this paper was supported in part by a grant from the Ford Foundation for the QUASAR project (grant number 890-0572). Any opinions expressed herein are those of the authors and do not necessarily reflect the views of the Ford Foundation.

2. Although space does not permit a thorough discussion of the varied purposes of external assessment, those who use information from such

tests to make instructional decisions should be mindful of the purposes for which the test was designed and use the information accordingly. Fairly complete, or otherwise interesting, treatments of different purposes and forms of externally developed or externally mandated assessment can be found in Airasian & Madaus (1972), Frechtling (1989), Linn (1983), Nitko (1989), Payne (1982), and Whitney (1989).

3. For a general discussion of the mismatch between commercial standardized tests and the goals of a "thinking curriculum," see Resnick & Resnick (1991) and R. G. Brown (1991). The relationship between external testing and mathematics education reform is discussed by Silver (1992).

4. QUASAR (Quantitative Understanding: Amplifying Student Achievement and Reasoning) is a research and development project aimed at the enrichment of the mathematics instructional program for students attending middle schools (grades 6–8) in economically disadvantaged communities (Silver, 1993). One of the project activities—in addition to providing technical assistance to schools and carefully monitoring program activities at the school and classroom levels—has been the design and administration of a collection of assessment tasks to measure growth in students' problem solving, reasoning, and communication.


## REFERENCES


Airasian, P. W. (1991). *Classroom assessment.* New York: McGraw-Hill.

Airasian, P. W., & Madaus, G. F. (1972). "Functional types of student evaluation." *Measurement and Guidance in Evaluation* 4, no. 4:221–233.

Bennett, A., & Foreman, L. (1990). *Visual mathematics: Course guide.* Portland, OR: Math Learning Center.

Borko, H., & Shavelson, R. J. (1990). "Teacher decision making." In B. F. Jones & L. Idol (eds.), *Dimensions of thinking and cognitive instruction,* pp. 311–346. Hillsdale, NJ: Lawrence Erlbaum Associates.

British Columbia Ministry of Education. (1990). *Tool box and handbook of assessment tools for process evaluation in mathematics.* Victoria, BC: Author.

Brown, C. A., Carpenter, T. P., Kouba, V. L., Lindquist, M. M., Silver, E. A., & Swafford, J. O. (1988). "Secondary school results for the fourth NAEP mathematics assessment: Algebra, geometry, mathematical methods, and attitudes." *Mathematics Teacher* 81, no. 5:337–351.

Brown, C. A., & Silver, E. A. (1989). "Data organization and interpretation." In M. M. Lindquist (ed.), *Results from the fourth mathematics assessment of the national assessment of educational progress*, pp. 28–34. Reston, VA: National Council of Teachers of Mathematics.

Brown, R. G. (1991). *Schools of thought: How the politics of literacy shape thinking in the classroom.* San Francisco: Jossey-Bass.

California Mathematics Council. (1986). "Standardized tests and the California mathematics curriculum: Where do we stand?" Unpublished report. (Available from L. Winters, California Mathematics Council, 18849 Clearbrook Street, Northridge, CA 91326.)

California State Department of Education. (1989). *A question of thinking: A first look at students' performance on open-ended questions in mathematics.* Sacramento: Author.

Carlsen, W. S. (1991). "Questioning in classrooms: A sociolinguistic perspective." *Review of Educational Research* 61:157–178.

Cazden, C. B. (1986). "Classroom discourse." In M. C. Wittrock (ed.), *Handbook of research on teaching,* pp. 432–463. New York: Macmillan.

Charles, R., Lester, F., & O'Daffer, P. (1987). *How to evaluate progress in problem solving.* Reston, VA: National Council of Teachers of Mathematics.

Clarke, D. J. (1988). *Assessment alternatives in mathematics.* Canberra, Australia: Curriculum Development Centre.

Collins, A. (1990). "Portfolios for assessing student learning in science: A new name for a familiar idea?" In A. B. Champagne, B. E. Lovitts, & B. J. Calinger (eds.), *Assessment in the service of instruction,* pp. 157–166. Washington, DC: American Association for the Advancement of Science.

Darling-Hammond, L., & Wise, A. E. (1985). "Beyond standardization: State standards and school improvement." *The Elementary School Journal* 85:315–336.

Dillon, J. T. (1985). "Using questions to foil discussion." *Teaching and Teacher Education* 1:109–121.

Dorr-Bremme, D. W., & Herman, J. (1986). *Assessing school achievement: A profile of classroom practices.* Los Angeles: Center for the Study of Evaluation, UCLA Graduate School of Education.

Dossey, J. A., Mullis, I. V. S., Lindquist, M. M., & Chambers, D. L. (1988). *The mathematics report card: Are we measuring up? Trends and achievement based on the 1986 national assessment.* Princeton, NJ: Educational Testing Service.

Doyle, W. (1986). "Classroom organization and management." In M. C. Wittrock (ed.), *Handbook of research on teaching,* 3rd ed., pp. 392–431. New York: Macmillan.

Elton, L. R. B. (1982). "Assessment for learning." In D. Bligh (ed.), *Professionalism and flexibility for learning.* Guildford, Surrey, England: Society for Research into Higher Education.

Erlwanger, S. H. (1973). "Benny's conception of rules and answers in IPI mathematics." *Journal of Children's Mathematical Behavior* 1:7–26.

Fennessy, D. (1982, July). "Primary teachers' assessment practices: Some implications for teacher training." Paper presented at the twelfth annual conference of the South Pacific Association for Teacher Education, Frankston, Victoria, Australia. (ERIC Document Reproduction Service No. ED 229 346.)

Fleming, M., & Chambers, B. (1983). "Teacher-made tests: Windows on the classroom." In W. E. Hathaway (ed.), *Testing in the schools: New directions for testing and measurement*, pp. 29–38. San Francisco: Jossey-Bass.

Frechtling, J. A. (1989). "Administrative uses of school testing programs." In R. L. Linn (ed.), *Educational measurement*, 3rd ed., pp. 475–483. New York: American Council on Education/Macmillan.

Glaser, R. (1986). "The integration of instruction and testing." In *The redesign of testing for the twenty-first century: Proceedings of the 1985 ETS invitational conference*, pp. 45–58. Princeton, NJ: Educational Testing Service.

———. (1988). "Cognitive and environmental perspectives on assessing achievement." In *Assessment in the service of learning: Proceedings of the 1987 ETS invitational conference*, pp. 37–43. Princeton, NJ: Educational Testing Service.

Green, J. L. (1983). "Research on teaching as a linguistic process: A state of the art." In E. W. Gordon (ed.), *Review of research in education*, vol. 10, pp. 152–252. Washington, DC: American Educational Research Association.

Gullickson, A. R. (1986). "Teacher education and teacher-perceived needs in educational measurement and evaluation." *Journal of Educational Measurement* 45:347–354.

———, & Ellwein, M. C. (1985). "Post hoc analysis of teacher-made tests: The goodness of fit between prescription and practice." *Educational Measurement: Issues and Practice* 4, no. 1:15–18.

Haertel, E. (1986, April). "Choosing and using classroom tests: Teachers' perspectives on assessment." Paper presented at the annual meeting of the American Educational Research Association, San Francisco.

Hill, S. A. (1991, April). "The vision: World class mathematics education for achieving the nation's goals." Opening remarks presented at the National Summit on Mathematics Assessment, Washington, DC.

Husen, T. (1983). "Are standards in U.S. schools really lagging behind those in other countries?" *Phi Delta Kappan* 64:455–461.

Kellaghan, T., Madaus, G. F., & Airasian, P. W. (1982). *The effects of standardized testing*. Boston: Kluwer-Nijhoff.

Kenney, P. A., & Silver, E. A. (1993). "Student self-assessment in mathematics." In N. Webb (ed.), *Assessment in the mathematics classroom, K–12: 1993 yearbook of the National Council of Teachers of Mathematics*, pp. 229–238. Reston, VA: National Council of Teachers of Mathematics.

Koretz, D., Klein, S., McCaffrey, D., & Stecher, B. (1993). *Interim report: The reliability of Vermont portfolio scores in the 1992–93 school year.* (CSE Technical Report 370). Los Angeles: National Center for Research on Evaluation, Standards, and Student Testing (CRESST), University of California, Los Angeles.

Kouba, V. L., Brown, C. A., Carpenter, T. P., Lindquist, M. M., Silver, E. A., & Swafford, J. O. (1988). "Results of the fourth NAEP assessment of mathematics: Measurement, geometry, data interpretation, attitudes, and other topics." *Arithmetic Teacher* 35, no. 9:10–16.

Lapointe, A. E., Mead, N. A., & Phillips, G. W. (1989). *A world of differences: An international assessment of mathematics and science.* Princeton, NJ: Educational Testing Service.

LeMahieu, P., & Leinhardt, G. (1985). "Overlap: Influencing what's taught: A process model of teachers' content selection." *Journal of Classroom Interaction* 21:2–11.

Lindquist, M. M. (ed.). (1989). *Results from the fourth mathematics assessment of the National Assessment of Educational Progress.* Reston, VA: National Council of Teachers of Mathematics.

Linn, R. L. (1983). "Testing and instruction: Links and distinctions." *Journal of Educational Measurement* 20, no. 2:179–189.

Madaus, G. F., West, M. M., Harmon, M. C., Lomax, R. G., & Viator, K. T. (1992). *The influence of testing on teaching math and science in grades 4–12.* Chestnut Hill, MA: Center for the Study of Testing, Evaluation, and Education Policy, Boston College.

McKnight, C. C., Crosswhite, F. J., Dossey, J. A., Kifer, E., Swafford, J. O., Travers, K. J., & Cooney, T. J. (1987). *The underachieving curriculum: Assessing U. S. school mathematics from an international perspective.* Champagne, IL: Stipes Publishing Company.

Mitzel, H. E. (1960). "Teacher effectiveness." In C. W. Harris (ed.), *Encyclopedia of educational research.* New York: Macmillan.

Mullis, I. V. S. (1990). *The NAEP guide: A description of the content and methods of the 1990 and 1992 assessments.* Washington, DC: U.S. Department of Education, Office of Educational Research and Development, National Center for Education Statistics.

———, Dossey, J. A., Owen, E. H., & Phillips, G. W. (1991). *The state of mathematics achievement.* Washington, DC: U.S. Department of Education, Office of Educational Research and Improvement, National Center for Education Statistics.

Mullis, I. V. S., Dossey, J. A., Owen, E. H., & Phillips, G. W. (1993). *NAEP 1992 mathematics report card for the nation and states.* Washington, DC: U.S. Department of Education, Office of Educational Research and Improvement, National Center for Education Statistics.

Mumme, J. (1990). *Portfolio assessment in mathematics.* Santa Barbara: California Mathematics Project, University of California, Santa Barbara.

National Assessment Governing Board. (1991). *The levels of mathematics achievement: Initial performance standards for the 1990 NAEP mathematics assessment.* Washington, DC: Author.

National Council of Teachers of Mathematics. (1989). *Curriculum and evaluation standards for school mathematics.* Reston, VA: Author.

———. (1991). *Professional standards for teaching mathematics.* Reston, VA: Author.

National Institute of Education. (1979). *Testing, teaching and learning: A report of a conference on research on testing, August 17–26, 1979.* Washington, DC: Author.

National Research Council. (1989). *Everybody counts: A report to the nation on the future of mathematics education.* Washington, DC: National Academy of Sciences.

Natriello, G. (1987). "The impact of evaluation processes on students." *Educational Psychologist* 22:155–175.

Nicholls, J. G., Cobb, P., Yackel, E., Wood, T., & Wheatley, G. (1990). "Students' theories about mathematics and their mathematical knowledge: Multiple dimensions of assessment." In G. Kulm (ed.), *Assessing higher-order thinking in mathematics,* pp. 137–154. Washington DC: American Association for the Advancement of Science.

Nitko, A. J. (1989). "Designing tests that are integrated with instruction." In R. L. Linn (ed.), *Educational measurement,* 3rd ed., pp. 447–474. New York: American Council on Education/Macmillan.

Pandey, T. (1991). *A sampler of mathematics assessment.* Sacramento: California Department of Education.

Payne, D. A. (1982). "Measurement in education." In H. E. Mitzel (ed.), *Encyclopedia of Educational Research,* 5th ed.: vol. 3. *Learning to Rehabilitation,* pp. 1182–1190. New York: The Free Press.

Peck, D. M., Jencks, S. M., & Connell, M. L. (1989). "Improving instruction through observation." *Arithmetic Teacher* 37, no. 3:15–17.

Resnick, L. B., & Resnick, D. P. (1991). "Assessing the thinking curriculum: New tools for educational reform." In B. R. Gifford & M. C. O'Connor (eds.), *Changing assessments: Alternative views of aptitude, achievement and instruction,* pp. 37–75. Boston: Kluwer Academic Publishers.

Romberg, T. A., Wilson, L., & Khaketla, M. (1989). *An examination of six standard mathematics tests for grade eight.* Madison, WI: Center for Education Research.

Romberg, T. A., Zarinnia, E. A., & Williams, S. R. (1989). *The influence of mandated testing on mathematics instruction: Grade 8 teachers' perceptions.* Madison, WI: National Center for Research in Mathematical Sciences Education.

Rose, B. (1989). "Writing and mathematics: Theory and practice." In P. Connolly & T. Vilardi (eds.), *Writing to learn science and mathematics,* pp. 15–30. New York: Teachers College Press, Columbia University.

Rosenshine, B., & Furst, N. (1973). "The use of direct observation to study teaching." In R. M. W. Travers (ed.), *Second handbook of research on teaching,* pp. 122–183. Chicago: Rand McNally.

Rotberg, I. C. (1990). "Resources and reality: The participation of minorities in science and engineering education." *Phi Delta Kappan* 71:672–679.

———. (1991). "Myths in international comparisons of science and mathematics achievement." *The Bridge* 2, no. 3:3–10.

Salmon-Cox, L. (1981). "Teachers and standardized achievement tests: What's really happening?" *Phi Delta Kappan* 62:631–634.

Science Research Associates. (1979). *SRA achievement series' users guide.* Chicago: Author.

Shavelson, R., & Stern, P. (1981). "Research on teachers' pedagogical judgments, plans, and behavior." *Review of Educational Research* 51:455–498.

Shell Centre for Mathematical Education. (1984). *Problems with patterns and numbers.* Nottingham, United Kingdom: University of Nottingham.

Shepard, L. A. (1989). "Why we need better assessments." *Educational Leadership* 46, no. 7:4–9.

Silver, E. A. (1993, March). *Quantitative understanding: Amplifying student achievement and reasoning.* Pittsburgh, PA: Learning Research and Development Center.

———. (1992). "Assessment and mathematics education reform in the United States." *International Journal of Educational Research* 17, no. 5:489–502.

———, & Adams, V. M. (1987). "Using open-ended problems." *Arithmetic Teacher* 34, no. 9:34–35.

———, Kenney, P. A., & Salmon-Cox, L. (1992). "The content and curricular validity of the 1990 NAEP mathematics items: A retrospective analysis." In National Academy of Education, *Assessing student achievement in the states: Background studies,* pp. 157–218. Stanford, CA: National Academy of Education.

———, & Kilpatrick, J. (1988). "Testing mathematical problem solving." In R. I. Charles & E. A. Silver (eds.), *The teaching and assessing of mathematical problem solving,* pp. 178–186. Reston, VA: National Council of Teachers of Mathematics.

————, Kilpatrick, J., & Schlesinger, B. (1990). *Thinking through mathematics*. New York: College Entrance Examination Board.

————, & Lane, S. (1993). "Assessment in the context of mathematics instruction reform: The design of assessment in the QUASAR project." In M. Niss (ed.), *Cases of assessment in mathematics education: An ICME study*, pp. 59–69. Dordrecht: Kluwer Academic Publishers.

————, & Mamona, J. (1990). "Stimulating problem posing in mathematics instruction through open problems and 'what-if-nots.' " In G. W. Blume & M. K. Heid (eds.), *Implementing new curriculum and evaluation standards: 1990 yearbook*, pp. 1–7. University Park: Pennsylvania Council of Teachers of Mathematics.

Smith, M. L. (1991). "Put to the test: The effects of external testing on teachers." *Educational Researcher* 20, no. 5:8–11.

Snow, R. E., & Lohman, D. F. (1989). "Implications of cognitive psychology for educational measurement." In R. L. Linn (ed.), *Educational measurement*, 3rd ed., pp. 263–331. New York: American Council on Education/Macmillan.

Souviney, R., Britt, M., Gargiulo, S., & Hughes, P. (1990). *Mathematical investigations: A series of situational lessons (Book one and book two)*. Palo Alto, CA: Dale Seymour Publications.

Stenmark, J. K. (1989). *Assessment alternatives in mathematics: An overview of assessment techniques that promote learning*. Berkeley: University of California Press.

Stephens, M., & Money, R. (1993). "New developments in senior secondary assessment in Australia." In M. Niss (ed.), *Cases of assessment in mathematics education: An ICME study*, pp. 155–171. Dordrecht: Kluwer Academic Publishers.

Stiggins, R. J. (1985). "Improving assessment where it means the most: In the classroom." *Educational Leadership* 43, no. 2:69–74.

————, & Bridgeford, N. J. (1985). "The ecology of classroom assessment." *Journal of Educational Measurement* 22, no. 4:271–286.

Trowell, J. (ed.) (1990). *Projects to enrich school mathematics, Level 1*. Reston, VA: National Council of Teachers of Mathematics.

Tyson-Bernstein, H. (1988). "America's textbook fiasco: A conspiracy of good intentions." *American Educator* 12:20–27, 39.

Vermont Department of Education. (1991). *Looking beyond "the answer": Vermont's mathematics portfolio assessment project*. Montpelier: Author.

Whitney, D. R. (1989). "Educational admissions and placement." In R. L. Linn (ed.), *Educational measurement*, 3rd ed., pp. 515–525. New York: American Council on Education/Macmillan.

# 4 ❖ Assessment: No Change without Problems

*Jan de Lange*

## CHANGING MATHEMATICS EDUCATION

Mathematics education is changing rapidly in a number of countries. In several—The Netherlands, Denmark, Australia—these changes began to take place in the 1980s. Other nations are currently in the process of reforming mathematics education. One of the most visible dialogues regarding change has been taking place in the United States, where the Mathematical Sciences Education Board (1990) has advocated restructuring the entire mathematics curriculum in terms of the following changes in the context of mathematics education.

- *Changes in the need for mathematics.* As the economy adapts to information-age needs, workers in every sector must learn to interpret intelligent, computer-controlled processes. Most jobs now require analytical rather than merely mechanical skills, so most students need more mathematics in school as preparation for routine jobs. . . . Similarly, the extensive use of graphical, financial, and statistical data in daily newspapers and in public policy discussions suggests a higher standard of quantitative literacy for the necessary duties of citizenship.
- *Changes in mathematics and how it is used.* In the past quarter of a century, significant changes have occurred in the nature of mathematics and the way it is used. In part, it is because of the nature and rapidity of these changes that the social constructivist philosophy of mathematics has emerged. Not only has much new mathematics been discovered but also the types and variety of problems to which mathematics is applied have grown at an unprecedented rate. Most visible, of course, has been the development of computers and the explosive growth of computer applications. Most of these applications have required the

development of new mathematics in areas in which this was not feasible before the advent of computers (Geoffrey Howson, personal communication).

- *Changes in the role of technology.* Computers and calculators have changed the world of mathematics profoundly. They have affected not only what mathematics is important but also how mathematics is done (Rheinboldt, 1985). The changes in mathematics brought about by computers and calculators are so profound as to require readjustment in the balance and approach to virtually every topic in school mathematics.

- *Changes in American society.* The changing demographics of the country and the changing demands of the workplace are not reflected in similar changes in school mathematics (MSEB, 1989). In the early years of the next century, when today's school children will enter the workforce, most jobs will require greater mathematical skills (Johnston & Packer, 1987). At the same time, white men—the traditional base of mathematically trained workers in the United States—will represent a significantly smaller fraction of new workers (Oaxaca & Reynolds, 1988). Society's need for an approach to mathematics education that ensures achievement across the demographic spectrum is both compelling and urgent.

- *Changes in understanding of how students learn.* Learning is not a process of passively absorbing information and storing it in easily retrievable fragments as a result of repeated practice and reinforcement. Instead, students approach each new task with some prior knowledge, assimilate new information, and construct their own meanings (Resnick, 1987). This constructivist, active view of learning is obviously consistent with the social or cultural view of mathematics and must be reflected in the way mathematics is taught.

- *Changes in international competitiveness.* Just as recognition of the global economy is emerging as a dominant force in American society, many recent reports have shown that U.S. students do not measure up in their mathematical accomplishments to students in other countries (e.g., Lapointe, Mead, & Phillips, 1989; McKnight et al., 1987; Stevenson, Lee, & Stigler, 1986; Stigler & Perry, 1988). The implications of such data for employers is that the American workforce will not be competitive with workers from other countries.

These points make the argument that a complete redesign of the content of school mathematics and the way it is taught are urgent.

Changing global conditions have led to changing goals in the schools. In The Netherlands, for instance, national educational goals, for the majority of the children, are

1. To become an intelligent citizen (mathematical literacy);
2. To prepare for the workplace and for future education;
3. To understand mathematics as a discipline.

Such goals resemble closely those articulated by the British Committee of Inquiry into the Teaching of Mathematics in Schools (Cockcroft, 1982) as responsibilities of the teacher:

- Enabling each pupil to develop, within his [and her] capabilities, the mathematical skills and understanding required for adult life, for employment, and for further study and training.
- Providing each pupil with such mathematics as may be needed for the study of other subjects.
- Helping each pupil to develop so far as it is possible [an] appreciation and enjoyment of mathematics itself and [a] realization of the role it has played and will continue to play both in the development of science and technology and of our civilization.
- Above all, making each pupil aware that mathematics provides . . . a powerful means of communication.

A fourth set of goals addressing change in mathematics education was prepared by the Commission on Standards for School Mathematics of the National Council of Teachers of Mathematics in 1989. In its report, *Curriculum and Evaluation Standards for School Mathematics*, NCTM lists four societal goals and five goals for students (NCTM, 1989).

The four general societal goals for mathematics education are

1. *Mathematically literate workers.* The technologically demanding workplace of today and the future will require mathematical understanding and the ability to formulate and solve complex problems, often with others.

2. *Lifelong learning.* Most workers will change jobs frequently and so need flexibility and problem-solving ability to enable them to explore, create, accommodate to changed conditions, and actively create new knowledge over the course of their lives.

3. *Opportunity for all.* Because mathematics has become "a critical filter for employment and full participation in our society," it must be made accessible to all students, not just the white males, the group that currently studies the most advanced mathematics.

4. *An informed electorate.* Because of the increasingly technical and complex nature of current issues, participation by citizens requires technical knowledge and understanding, especially skills in reading and interpreting complex information. (NCTM, 1989, pp. 3–5)

Then asserting that educational goals for students "must reflect the importance of mathematical literacy," NCTM proposes five general goals for students:

1. *Learning to value mathematics.* Understanding its evolution and its role in society and the sciences.

2. *Becoming confident of one's own ability.* Coming to trust one's own mathematical thinking and having the ability to make sense of situations and solve problems.

3. *Becoming a mathematical problem solver.* This is essential to becoming a productive citizen and requires experience in solving a variety of extended and nonroutine problems.

4. *Learning to communicate mathematically.* Learning the signs, symbols, and terms of mathematics.

5. *Learning to reason mathematically.* Making conjectures, gathering evidence, and building mathematical arguments.

These goals imply that students should be exposed to numerous and varied interrelated experiences that encourage them to value the mathematical enterprise, to develop mathematical habits of mind, and to understand and appreciate the role of mathematics in human affairs; that they are encouraged to explore, to guess, and even to make errors so that they gain confidence in their ability to solve complex problems; that they read, write, and discuss mathematics; and that they conjecture, test, and build arguments about a conjecture's validity. . . . The opportunity for all students to experience these components of mathematical training is at the heart of our vision of a quality mathematics program. The curriculum should be permeated with these goals and experiences such that they become commonplace in the lives of students. (NCTM, 1989, pp. 5-6)

Each of the five goals statements reflects a shift away from traditional practice. Traditional skills are subsumed under more general goals for problem solving, communication, and the development of a critical attitude.

*Changing Theories*

At the same time that the goals of mathematics education are changing, we are also witnessing the evolution of new theories for the learning and teaching of mathematics. Romberg (1991) points out that these sets of goals all implicitly reflect a social constructivist philosophy of mathematics. Galbraith (1993) compares the conventional and constructivist paradigms, as paraphrased below:

- There exists a simple reality that is realized in universal laws and can be verified by objective observation (Conventional), as opposed to a series of multiple constructed realities, where truth is relative (Constructivism).
- Facts and values are independent in the conventional view, but interdependent in the constructivist view.
- Problem solutions have widespread application across contexts (Conventional); problem solutions have only local applicability (Constructivism).
- Phenomena have no meaning except in the context for which the construction occurred (Constructivism). (pp. 73–74)

The constructivist is prepared to examine a set of results and consider their possible application to other situations given the contextual features of both. Transferability, rather than generalizability, characterizes this aspect of the search for consensus.

At the Freudenthal Institute, the "theory for realistic mathematics education" evolved after twenty years of developmental research that in several important respects correlates with the constructivist approach (see de Lange, 1987; Freudenthal, 1983, 1991; Gravemeijer, van den Heuvel-Panhuizen, and Streefland, 1990; Treffers, 1987). There are, however, some differences.

The social constructivist theory is in the first place a theory of learning in general, whereas realistic mathematics theory is a theory of learning and instruction that evolved only in mathematics. One of the key components of realistic mathematics education is that students reconstruct or reinvent mathematical ideas and concepts through exposure to a variety of "real-world" problems and situations.

This process takes place by means of progressive schematization and horizontal and vertical mathematization. The students are given opportunities to establish their own pace in the concept-building process. At some point, abstraction, formalization, and generalization take place, although this may not occur for all students. The question, for instance, of how far we can be successful within mathematics if our students master "only" the skill of transferability, instead of generalizability, is still open for discussion.

### Changing Content

It is not only goals and teaching and learning theories of mathematics education that have changed. New subjects are slowly and sometimes cautiously introduced into curricula—a prominent example is discrete mathematics, and there seems to be a revival of geometry. Some of these subjects take their place in the curriculum because new technologies have opened new possibilities. The computer has had some (limited) impact on the teaching of mathematics, but future development might have more visible effects. A graphic calculator with a computer algebra system would outdate both personal computers and graphic calculators as we now know them. Also, if interactive CD enters the consumer market, it will in all likelihood become an important tool in education as well.

Apart from these external factors, internal factors are operating to change the content of school mathematics. We mentioned the revival of geometry that, offered with new insights and in a broader context, gives it different content. Other domains that would expand the curriculum, including the central place of calculus, the emphasis on fractions and percentages, the role of logarithms, have been discussed in the last decade with mixed results. Finally, new insights into how children learn and what didactical tools we possess to enable children to understand better certain mathematical tools are important issues. Changing learning theories can definitely lead to new content subjects.

### Changing Assessment

There seems to be a lot of truth in Galbraith's conclusion (1993) that we need to confront inherent contradictions that exist when constructivism drives curriculum design and knowledge construction, but that positivistic remnants of the conventional paradigm drive the assessment process. In The Netherlands, this distinction confronted us with a paradox. Many teachers and researchers react with, "I like the way you have embedded your math education in a rich context, but I will wait for the national standardized test to see if it's been successful." Popper (1968) and Phillips (1987) have argued that a theory can be tested only in terms of its own tenets. This means that the constructivist or realistic mathematics education of teaching and learning can be evaluated only by assessment procedures derived from the same principle. It also means that assessment procedures should do justice to the goals of the curriculum and to the students. Context-independent generalized testing is unjust in such a case (most of the time, the context will also include the real world of mathematics itself, at least in the realistic mathematics education approach). Therefore, an essential question is, Does assessment reflect the theory of instruction and learning represented by the curriculum?

Not only have new notions about learning influenced the ideas about "authentic" assessment, but the new goals, emphasizing reasoning, communication, and the development of a critical attitude, will have an impact. Popularly associated with "higher order" thinking, these skills were seldom or never present in traditional education and assessment. The change toward a "thinking" curriculum forces us to focus on "thinking" assessment as well.

In the next four sections of this chapter, therefore, we will examine levels in assessment, the role of content, necessary and sufficient information, and different test formats. In the final section, we consider briefly a number of issues that bear directly on the design of assessments for use in the learning and teaching of authentic mathematics.


## LEVELS IN ASSESSMENT


Most instruction in mathematics education has focused on learning to name concepts and objects and to follow specific procedures. The result, as Bodin (1993) points out, is that a student can solve a given equation without being able to express the steps taken or to justify the results without knowing which type of problem it is connected to, or without being able to use it as a tool in another situation. As an example, Bodin observed children who were able to solve the following equation,

$$7x - 3 = 13x + 15$$

but who were unable to answer the question:

Is 10 a solution to the equation $7x - 3 = 13x + 15$?

Here we notice different levels of "knowledge." The equation can be solved simply by following a procedure, but the latter question requires judgment.

In a global economy, the emphasis for all students must shift from following rote procedures to the development of the higher order thinking skills. L. Resnick (1987) listed salient features of higher order skills, many of which are in stark contrast to the mathematics criteria that prevail in many schools. She noted that higher order thinking skills tend to be complex, their total path not "visible" (mentally speaking) from any single vantage point. Furthermore, higher order thinking involves the application of multiple criteria that sometimes conflict with one another.

Experiences with higher order thinking in mathematics and its assessment have also been described in some detail by de Lange (1987), who stressed the process-versus-product character of the new curriculum. During experiments in The Netherlands over the last decade, it became clear that the mathematics in the new curriculum is nonalgorithmic, has multiple solutions, and involves uncertainty and a need for interpretation. Thus, one of our major challenges is to find structure in apparent disorder; we need to carry out considerable work in the kinds of elaborations and judgments required to reinforce the place of higher order thinking in the new curriculum.

Further, we need to address the problem at the different cognitive levels in both instruction and assessment. For such needs, we identify three cognitive levels, which, although somewhat arbitrary, are sufficient and reflect the decade-long experience we had with our research on the content and implementation of the new mathematics curriculum in The Netherlands. To describe these three levels, it is first necessary to define the guiding principles and goals for assessment that we employed.

*New Principles and Goals for Assessment*

In The Netherlands in 1987, we formulated five principles that have guided our assessment work:

- The first and primary purpose of testing is to improve learning and teaching.

- Methods of assessment should enable the students to demonstrate what they know rather than what they do not know.
- Assessment should operationalize all of the goals of mathematics education.
- The quality of mathematics assessment is not determined by its accessibility to objective scoring.
- The assessment tools should be practical (de Lange, 1987).

*The First and Primary Purpose of Testing is to Improve Learning and Teaching.* This first principle is easily underestimated in the teaching-learning process. All too frequently we think of testing as an end-of-the-unit or end-of-the-course activity whose primary purpose is as a basis for assigning course grades. A properly designed test or task should not only motivate students by providing them with short-term goals toward which they work, but also by providing them with feedback concerning the learning process. Furthermore, more complex learning results, such as levels of understanding, application, and interpretation, are likely to be retained longer and have greater transfer value than results at the knowledge level. This means that we should include measures of these more complex learning results in our tests. In this way we provide students practice in reinforcing the comprehension, skills, applications, and interpretations that we are attempting to develop.

*Methods of Assessment Should Enable Students to Demonstrate What They Know Rather Than What They Do Not Know.* This principle—sometimes referred to as *positive testing*—is borrowed from Cockcroft (1982). Most traditional testing consists of checking what the students do not know; students are given a specific problem that has, in most cases, a single solution. If the student does not know how to solve the problem, there is usually no way to gauge what he or she does know. One result may be that the student loses confidence—an effect not conducive to promoting "the development of the talents of all people" (MSEB, 1991).

*Assessment Should Operationalize All of the Goals of Mathematics Education.* The fact that tasks that operationalize "higher order thinking skills" are difficult to design and score should be no reason to restrict ourselves to the testing as usual. It is essential that we be able to test the capacity of students for mathematization, reflection, discussion of models, communication, creativity, generalization, and transfer. This means also that we are less interested in the product than in the process that leads to this product. The consequence is that we need a variety of effective assessment methods.

*The Quality of Mathematics Assessment Is Not Determined by Its Accessibility to Objective Scoring.* This principle is a very important one. In the first place, the quality of a test is frequently derived from the accessibility to mechanical or objective scoring—a problem endemic in the United States. It may be difficult to score more complex tasks but experience shows that at the same time the advantages are much greater than the perceived disadvantages. In the first place, in complex context problems the problems become much easier for the student to understand if he or she makes the problem his or her own and the answers show what the student is actually capable of doing. In traditional tests, we often cannot even tell whether the student understands the question fully, let alone find in the answer any indication of level of understanding. Second, the professional mathematician is not evaluated on the basis of tests but on the basis of his or her output. Finally, an important aspect of educating the mathematics education community is the development of new forms of assessment and of guidelines for judging all forms of assessment.

*The Assessment Tools Should be Practical.* What we mean by *practicality* is that assessment should have practical applications in the school. At a given school, a balanced package of assessment tools will be different from those used at another school because of physical limitations, differences in school culture, accessibility to outside resources, and other factors. We need also to bear in mind the demands made by assessment on the teacher.

### Lower Level Assessment

At the *lower level,* we are dealing primarily with traditional mathematics and traditional tests. This level concerns objects, definitions, technical skills, and standard algorithms. The following are typical examples of lower level problems:

- Solve the equation $7x - 3 = 13x + 15$.
- What is the average of 7, 12, 8, 14, 15, 9?
- Draw the graph of $y = -x^2 - 2x + 8$.
- Make a drawing that illustrates 1/4.

- Write 69% as a fraction.
- Line *m* is called the circle's ———. (Adapted from 1989 Illinois State Board of Education testing materials, with permission.)

Quite often, multiple-step problems from the real world are introduced at this level, although texts may treat them as standardized exercises with no real-problem meaning. The following illustrate this point:

- Christine borrowed $168 from the Friendly Finance Company. She had to pay 6% interest.
  —How much interest will she pay in one year? (NAEP, 1990)
  —We drove our car 170 miles and used 4 gallons of gasoline.
  —What was our mileage—that is, how many mpg?

It can be argued that these items require more processing. But, in fact, the solutions simply involve routine, sequential processing.

*Middle Level Assessment*

The *middle level* can be characterized by having students relate two or more concepts or procedures; thus, *making connections*, *integration*, and *problem solving* are terms often used to describe this level. It is more difficult to provide examples at this level from extant testing sources, although there have been good tests that operationalize the principles we have articulated. However, several examples from our work, and the new work of others, illustrate the possibilities presented by the new curricula:

- You have driven your car $2/3$ of the distance you want to cover and your tank is $1/4$ full. Do you have a problem? (fifth grade) (Streefland, in Gravemeijer et al., 1990)

- In pictures A and B that follow:
  —How many of the cartons will be left after the box is filled? (A)
  —How many cartons do we need for 81 children? (B) (van den Heuvel-Panhuizen & Gravemeijer, 1990). Used with permission.

---

Note: Test items cited in this chapter may be released, but unpublished.

**A**

**B**

■ You have a supply of $2 \times 1$ rectangles like this one:

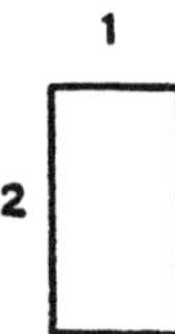


■ You can use these rectangles to make other rectangles that are 2 units deep and of whatever width you choose. For example, here are some $2 \times 3$ rectangles:

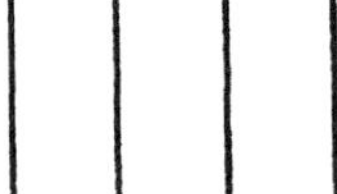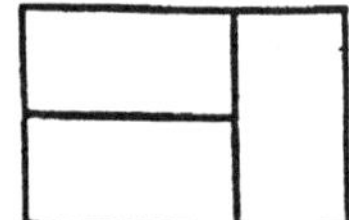

—Describe how many $2 \times n$ rectangles it is possible to make from $2 \times 1$ rectangles (where $n$ is a natural number). Justify your conclusion.

—Extend your solution to describe how many $3 \times 4$ rectangles can be made from $3 \times 1$ rectangles.

—Extend your solution further to describe how many $m \times n$ rectangles can be made from $m \times 1$ rectangles (where *m* and *n* are natural numbers). (Victoria Curriculum and Assessment Board, 1990. Used with permission.)

Another example from a new standardized test in The Netherlands:

■ Here you see three squares. In each one, a border of width 1 cm has been made black.

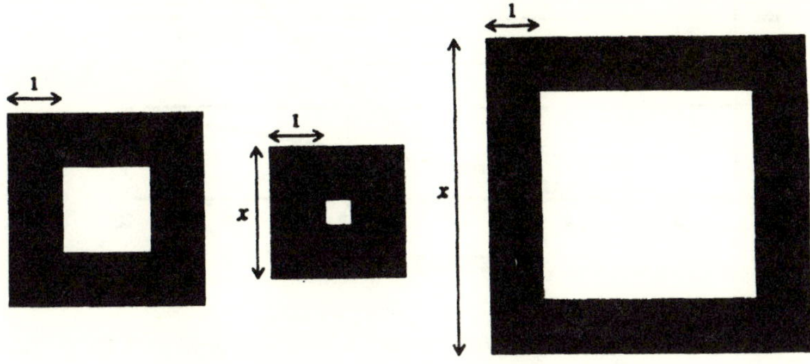

—Draw such a square with side $x$ = 5 cm.
—When the border has a width of 1 cm, the area ($A$) of the white inner part can be represented by the formula:

$A$ (white) = $(x - 2)^2$.

Check this for the square you have just drawn.

—Check whether or not the formula for $A$ (white) has a meaning in each of the following cases:

$x = \sqrt{3}; \quad x = \sqrt{5}; \quad x = 1,000,000.$

—Compute for which value of $x$ the area equals 400.
—For the black area the formula representing the area is

$A$ (black) = $4x - 4$.

Compute for which value of $x$ the area equals 400. The following graphs represent the formulas $A$ (white) and $B$ (black):

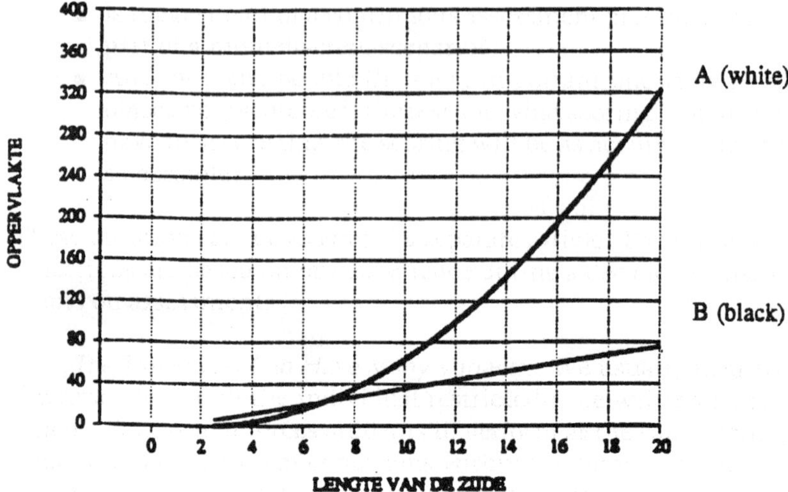



—Compute (in one decimal) the value(s) of $x$ for which the white and black areas are equal.

- Sylvia has computed that for $x = 7.4$, the area for the white and black part are equal. Explain why this cannot be the case. (W12–16, team, 1991. Used with permission.)

The examples—all taken from tests that are in use—clearly indicate features that do not belong on the lowest level. It is interesting to compare the two car fuel problems. In the first (we drove 170 miles and used 4 gallons of gasoline. How many mpg?), the students have already been trained to grab their calculator and to simply divide 170 by 4. The other problem (you have driven $^2/_3$ of the distance, and $^1/_4$ of a tank of fuel is left. Do you have a problem?) does not describe a certain strategy. The children left on their own must design their own strategy. As a side effect—but a very important one—the teacher will get valuable feedback on the level of understanding of the student. Actual solutions of the (primary school) students in figure 4.1 show remarkable strategies (Streefland, 1990. Used with permission).

The items about milk cartons aim at more complex activities of a problem-solving nature (van den Heuvel-Panhuizen & Gravemeijer, 1990). The true-to-life contexts not only help the children to grasp immediately the situation of the items, but they also offer the opportunity to sound out the children's abilities while avoiding the obstructions caused by formal notation. The question in the milk carton problem is: How many do not fit in the box?

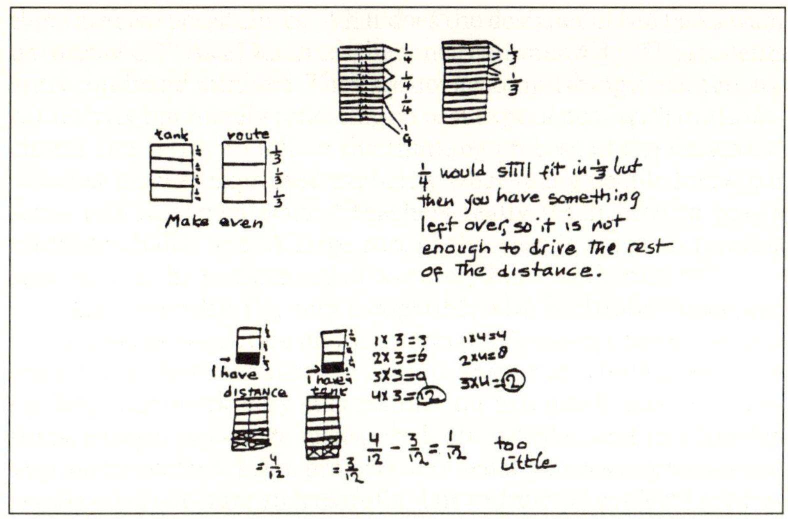

Figure 4.1   Children's strategies for solving the "2/3 distance, 1/4 tank of fuel" problem. Used with permission.

whereas the item on the carton shown with the 6 cups asks the following question: How many packages of chocolate milk are needed for 81 children? First of all, the arithmetic operation is not obvious; moreover, even a correct calculation of $81 \div 6$ does not directly yield an adequate answer. These kinds of questions give the teacher information about the children's informal knowledge, and solutions can be used to attune the teaching to the children's previous knowledge. In this way, tests become an instrument the teacher can use to improve learning.

A last example that might qualify for the middle level is the following:

- 200 women and 200 men were given a test on how to run a family. The maximum possible score was 16. The results are represented in a box plot as follows:

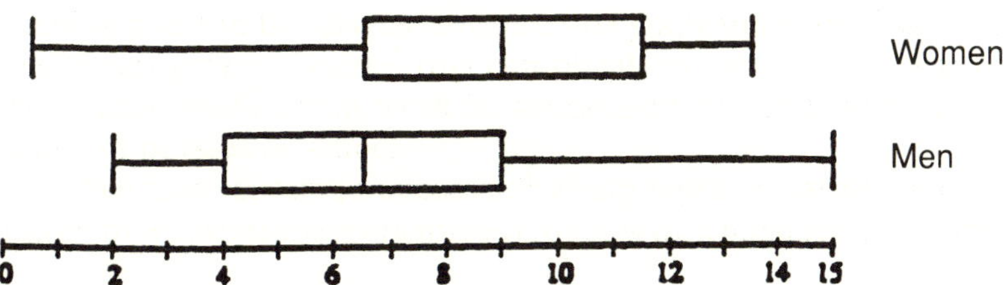

—Is it possible for the following box plot to represent the results of the test for all 400 participants? Explain your answer (de Lange, Burrill, Romberg, & van Reeuwijk, 1993).

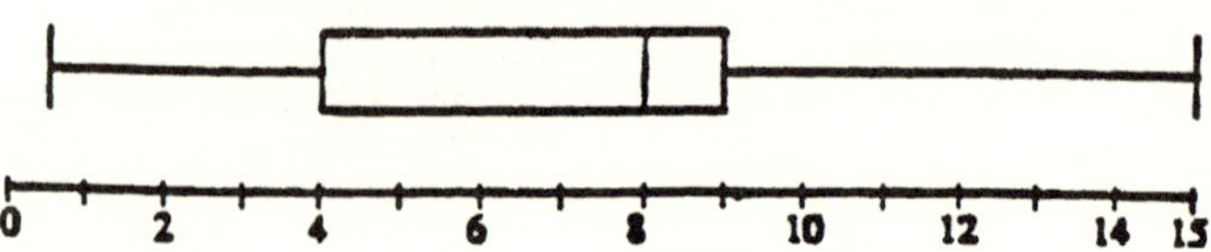

This last question demands some real understanding of the box plot. The students agree that this is a difficult question. And even though they often sense what is wrong, they do not always have a very clear way of thinking on paper; for example, two responses were

- No, it is impossible because its center, 50%, is in the wrong position. (This, of course, is true. But the student failed to further communicate his reasoning properly on paper. So we might consider qualifying this item for the highest level, because of the reasoning and communication skills necessary to solve the problem. There were students who thought they lacked adequate information.)
- No, it would not be possible because you would have to write out all the data, find the new median, and that means that the first and second quartiles would be different.

Or an even stronger response:

- There is no data to base this chart on available to me!

"Correct" answers sometimes cannot be given appropriate credit because of the lack of information on the reasoning:

- No, the first quartile is not correct, neither is the third quartile. A correct answer could be this: For women, 50% lies right of the 9, and for men 25%. This means that 100 + 50 = 150 persons lie on the right of 9. So it is impossible that the new box plot represents the results on the test.

*Higher Level Assessment*

It is even more difficult to describe *higher level* assessment than that at the middle level. This, of course, is partly because we are dealing with more complex material: mathematical thinking and reasoning, communication, critical attitude, interpretation, reflection, creativity, generalization, and mathematizing. We will highlight aspects of tests that operationalize some higher order thinking skills—at different school levels. A major component will be the "construction" by children that completes the problem.

Let us first look at some of the more "open" tests at primary school level. Especially when we are dealing with nonalgorithmic problems that relate to the student's real world, we also need to know the procedures the children use. Or, to put it even more strongly, we are more interested in the process than in the product—that is, the answer—because, of course, there might be multiple solutions. All of these arguments apply to the following test items. The first relates to the visit of a circus.

- The total admission costs for the children are $50. How much were the tickets? (van den Heuvel-Panhuizen & Gravemeijer, 1990. Used with permission.)

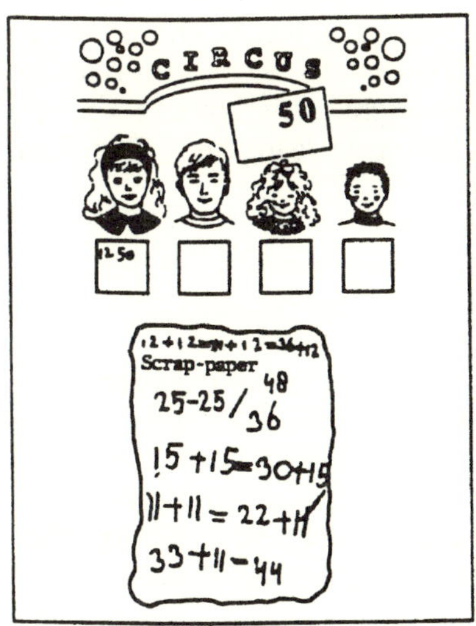


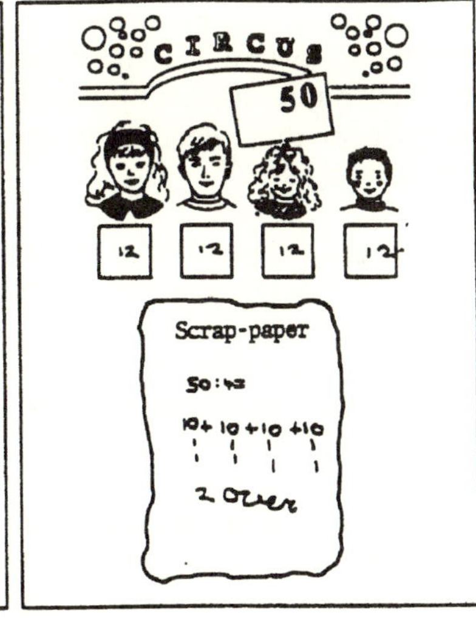

The first pupil tries to approximate the total amount of $50 as closely as possible, while the others apply a formal division, or a less formal distribution strategy. The item poses the following question:

—How many children weigh the same as this bear?

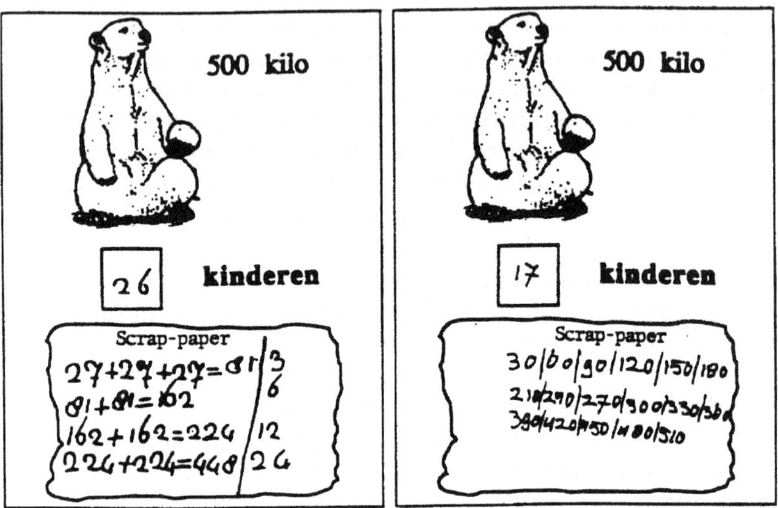


The bear item refers to the children's knowledge of measures. Only the weight of the polar bear is given. It is left to the pupils to determine how much a child generally weighs. Some children, like the first pupil, stick to their own weight; others prefer a round number, or they weigh precisely 30 or 25 kilograms. The third item is:

- Design as many sums as possible with the answer of 100 (workspace is provided on the answer page).

The objective of the third item is to elicit the capability of children for individual productions, their own responses. The child is asked to think, rather than to solve problems (Streefland, 1990; van den Brink, 1987). A simple way to estimate the scope of children's abilities is to ask them to produce an easy and a difficult sum. Due to this latitude in devising their own productions, children reveal not only what they are capable of, but also their manner of working.

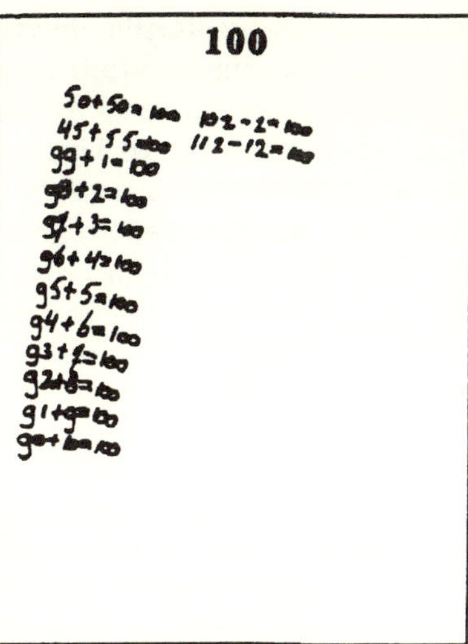

Some record only isolated sums, whereas others proceed systematically, for instance, by always changing the first term by one unit or by applying commutativity.

Another primary school item that might fit on the highest level is the following:

- Martin lives three miles from school and Alice five miles. How far apart do Martin and Alice live from each other?

This item might be seen as belonging to geometry. Or, perhaps it can be solved by just common sense reasoning. Or by visualizing. Multiple strategies are possible, at different levels. But it is almost certain that the students have never been purposely presented with an isomorphic exercise.

Approximately seventy teachers were interviewed about the appropriateness of this item. Some of their first reactions to the question were, "5 − 3 = 2, so it's a simple subtraction (lower level) and for that reason we don't like it as a test item." A second reaction was, "You can't say the proper answer because there is no one proper answer, and for that reason this is not a good test item." A third reaction was: "You can't give the proper answer because there is not one proper answer, and for that reason it's a good test item." A typical reaction, falling in this third category, was, "You can't tell it exactly, but you can say something. For instance, that Martin and

Alice cannot live farther away from each other than 8 kilometers, or no closer than 2 kilometers. You can show that with a nice picture."

Looking at it in this way makes it a rich item that offers many possibilities for different strategies reflecting the reasoning of the students. But the teachers' reactions clearly demonstrate that we have a long way to go if we want to implement this kind of question. In a group of teachers favoring development of more "realistic" mathematics education, only 17 percent offered arguments like the one just quoted. The majority of teachers using more traditional books (57 percent) thought the item was unfit because it did not have one single answer. This ambiguity—the lack of one single answer—was by far the most frequent argument for including the item on the part of the teachers who liked it. In this brief analysis, it is evident how difficult the process of change toward "new" modes of assessment will be (Gravemeijer et al., 1992).

There are many other techniques for encouraging students to "produce." The following example is a rather simple item:

- If no fish were caught, the number of fish will increase during the coming years. The graph shows a model of the growth in the number of fish.

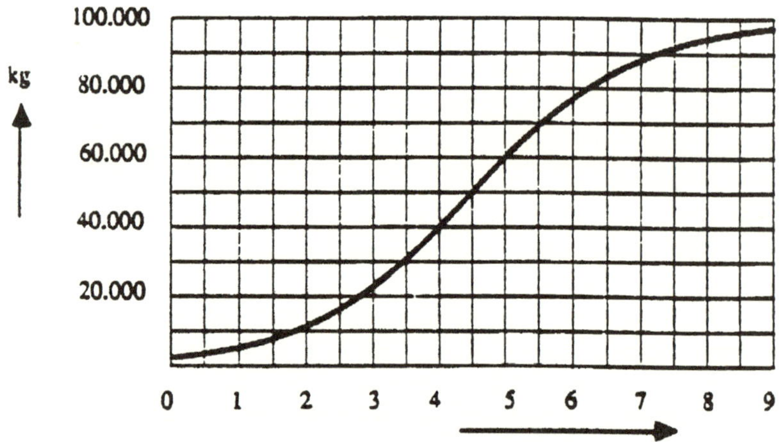


—Draw an increase diagram with intervals of a year, to start with the interval 1–2. The fish farmer will wait some years before he will be able to catch annually the

> same amount of fish as the first year; after every catch the number of fish increases again according to the graph.
> —What would you advise the fish farmer about the number of years he has to wait after planting the fish and the amount of fish that he will catch every year? Give convincing arguments (HAVO, National Examination, The Netherlands, experiment, 1990).

One should be aware, when looking at this exercise, that the students were not familiar with the "differentiation of functions," but they did know about the changes and rates of change of real phenomena in a discrete way; that is, they were not used to graphing the derivative of a function, but they were accustomed to using increase diagrams. So the first question was very straightforward, operationalizing only the lowest level.

The other question is a different story. It was both new to the students and new in its form on the national standardized test in The Netherlands. Communicating mathematics, drawing conclusions, finding convincing arguments are activities that all too often are not a part of mathematics tests and examinations. Many teachers were surprised by the richness of the question and did not know what to think of this development, although a few, those who identified the question with the new approach as indicated in the experiments, were prepared to some extent.

In some respects, the students seemed less surprised—if we are to judge by the results— although their answers showed a wide range of responses:

- I would wait for four years and then catch 20,000 kilos per year. You can't lose that way, man.
- If you wait till the end of the fifth year then you have a big harvest every year: 20,000 kg of fish; that's certainly not peanuts. If you can't wait that long, and start to catch one year earlier you can catch only 17,000 kg, and if you wait too long (one year) you can only catch 18,000 kg of fish. So you have the best results after waiting for five years. Be patient, wait those years. You won't regret it. (Van der Kooij, 1989. Used with permission.)

Generalization, adjustments of models, communication, reflection are only some of the characteristics of the following test; only a relevant part (relevant for the highest level) has been reproduced, because we will discuss this test later in more detail (see also de Lange, 1987).

A forester has a piece of land with 3,000 Christmas trees. He distinguishes three classes of length: S, M, and L. The small trees have just been planted and have no economic value; the medium trees are sold for $10 each and the large ones for $25. At the beginning of his first year as owner, his tree farm has 1,000 small, 1,000 medium, and 1,000 large trees. All of these grow uneventfully until just before Christmas.

From the experiences of colleagues, he knows approximately how much growth to expect per year:

40% of the small trees become medium
20% of the medium trees become large

(Here we omit some lower- and middle-level questions.)

The forester wonders what strategy for cutting and planting is the most profitable one. He considers several strategies:

  I.  Cut medium and large trees after one year and replant to return to the starting population of 1,000 of each kind.
 II.  Cut medium and large trees after two years and replant to return to the starting population of 1,000 of each kind.
III.  Cut after one year the large trees only (leaving 1,000) and replant that same number of small trees; repeat this the second year.

Cutting costs for one tree are $1 and planting costs $2.

- Decide which is the most profitable strategy over a two year period.
- Find the matrix that represents the general case (with n = length of classes):

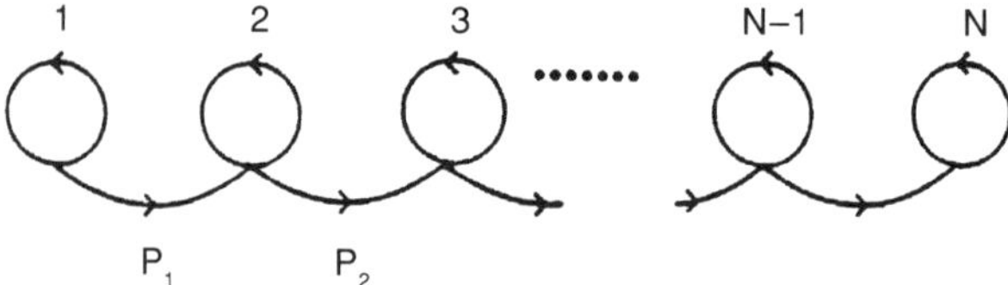


How can you conclude from the matrix whether or not it is possible to get the starting population back? What are the limitations of the model? What refinements would you like to suggest?

It is interesting to see that many students did well on this test—a point to be discussed later in this chapter. But we have to bear in mind that this test item was completed by students at home, so we feel it cannot be compared with other items discussed so far, which were for the most part meant for restricted-time written tests.

## THE ROLES OF THE CONTEXT

Problem-oriented mathematics education places mathematics in a context. In realistic mathematics education, the real world is used as a starting point for the development of mathematical concepts and ideas. According to Treffers and Goffree (1985), context problems in realistic curricula fulfill a number of functions:

- Concept formation: In the first phase of a course, they allow the students natural and motivating access to mathematics.
- Model formation: Context problems supply a firm basis for learning the formal operations, procedures, notations, rules, and they do this in conjunction with other models that function as important supports for thinking.
- Applicability: Context problems utilize reality as a source and domain of applications.
- Practice the exercise of specific abilities in applied situations.

In an earlier article (de Lange, 1979), distinctions were made among the uses of context in a way that fit with these four functions. One of the functions of context—and for realistic mathematics education the most important characteristic—is its use for concept formation, the conceptual mathematization process. This use of context presents problems in assessment that are somewhat different from the problems encountered in the other context classifications: that is, we usually will not introduce new concepts during a test, but *apply* new mathematical concepts in some way.

### Functionality

*No Context.*  This category hardly needs further elaboration. However, we cannot resist the temptation to include a recent example—from a standardized test from Poland (personal communication, W. Zawadowski)—that shows a complex task without context. The question that confronts us is this: At which level are we working here? Is this higher order because it is so complex, or is it lower order because it is repetitive?

What number is 75% of:
$$\frac{\sin^2 30° - (^1/_2)^2 \cdot (0.8)^{-1} + \sqrt{2.25}}{^{11}/_{20} + (^2/_3)^2 \cdot (\cos 60° + \tan 45°)^2}$$

*Camouflage Context.* The context in this situation is used only to "camouflage" or "dress up" the mathematical problem. Most of the so-called word problems and multistep problems from the NAEP (1990) are of this form. We refer, for instance, to the problem of Christine and the Friendly Finance Company. Similar problems would include

- The growth factor of a bacterium type is 6 (per time unit). At the moment there are 4 bacteria. Calculate the point in time when there will be 100 bacteria.
- The interest for a year on a savings account is 8%. $4,000 is deposited at time 0. At what point in time will this amount have increased to $5,000?

In this category, there are also the familiar types of items, such as the one that follows. The goal that should be operationalized with this item is: Identify, analyze, and solve the problem using algebraic equations, inequalities, and functions and their graphs (Adapted from 1989 Illinois State Board of Education testing materials, with permission, 1989). Although the problem obviously does not qualify as other than a lower level assessment item, in this form it is interesting to note that as presented, it looks quite different and certainly does not operationalize the desired goal:

- Which of the following number sentences could be used to solve the following problem? Bill weighed 107 pounds last summer. He lost 4 pounds, and then he gained 11 pounds. How much does he weigh now?

  a. $107 - (4 + 11) = A$
  b. $(107 - 4) + 11 = A$
  c. $(107 + 11) + 4 = A$
  d. $-4 + 11 = 107 + A$
  e. $(107 - 11) + 4 = A$

The issue is not that one has to solve the problem, but that one has to analyze notations that no intelligent person would use to solve the problem. It is an almost perfect example of an elaborated problem item that does not attain its desired goals. It may be very difficult to decide whether or not a problem has a camouflaged or elaborated context.

Of greater complexity is the role of context in the following test item. It seems, on the face of it, to fit into the middle level category, but the role of the context is deceptive. A model train track with a short and a long circuit is illustrated and the time (10 min) the train will take to complete the short track. The question is, How long will it take the train on the longer track? (van den Heuvel-Panhuizen & Gravemeijer, 1990).

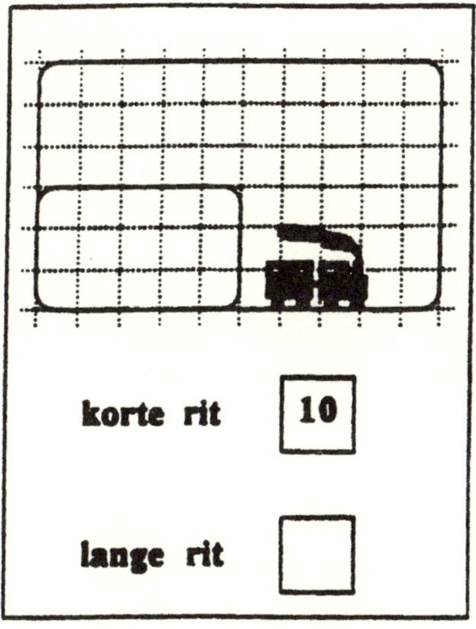


This is a simple question, an elaborated context, yet, neverthe-less, a middle-level test item because a thorough analysis of the problem is required and because of the built-in stratification; that is, the way the item is presented allows for solutions on several levels. However, the elaboration character has to be restricted to the train (could be a car, bus), but the map is the relevant context.

*Relevant and Essential Context.* We begin with a simple test for grade 1.

A    B

In A, the children are asked to buy "something" and circle the number that shows the money left in one's purse. Although there are several degrees of difficulty here, the choice gives indications of what children are capable of. Of course, preferences for a certain object play a part (relevant and essential context). Experiments have shown, however, that quite a few children make numerically similar choices on tests of this kind. (van den Heuvel-Panhuizen & Gravemeijer, 1990. Used with permission.)

Other types of problems show again the "own production" aspect of tests coupled with relevant context use. In B, the children are asked to make a program for a birthday party, or rather, to complete it; the starting time and the activities are already given. It is left to the children to determine how long each activity will take. The only thing that is predetermined is 45 minutes for the movie. As with most open items, this one allows a great many observation points: There must be progression of time, duration must be consistent with the activities planned, and finally, digital notation of time must at least be understood (van den Heuvel-Panhuizen & Gravemeijer, 1990).

As a final example of the relevant use of context, we have chosen a standardized test item that was revised in an effort to improve its efficacy:

> Among other things the quality of water in a swimming pool is judged on the basis of the amount of urea. Urea enters the water via perspiration and urine. It appears that the average daily increase in the amount of urea is 500 g per 1,000 visitors a day. The water must be refreshed in such a way that the statutory standard of 2 g urea per cubic meter ($m^3$) will not be exceeded.
>
> In the model, we make the assumption that 1,000 swimmers visit the pool, which has a volume of 1,000 $m^3$, daily. The

refreshing of the water takes place at night. For each daily visitor, 30 liters (l) of water will be refreshed. In our model, this means a refreshment of 30 m$^3$ (3% of the total).

The first day we start with 0 g of urea in the water. At the end of the day, the water contains 500 g urea. After refreshing, there will be an excess of 485 g urea at the beginning of the second day.

> —Show by calculation that the amount of urea is more than 955 g at the beginning of the third day.
> —In the course of which day will the statutory standard be exceeded?

A refreshment of 30 l per visitor is not sufficient. Suppose in the model, 200 l will be refreshed instead of 30 l. Let $U$ be the amount of urea at the beginning of a certain day. Show that the amount of urea is 0.8 + 400 at the beginning of the next day. In our model, we start again with 0 g urea at the beginning of the first day. The amount of urea $(U_n)$ at the beginning of the $n$th day can be calculated directly with the formula:

$$U_n = 2000 - 2500 \cdot (0.8)^n$$

> —Explain with the aid of this formula that at the beginning of each day the amount of urea will meet the requirements. In the course of the day the statutory standard can be exceeded.
> —On which day will this happen for the first time?
> (HAVO, National Examination, The Netherlands, 1991)

We are not often in a position to describe in detail how a rather complicated item like the last one is constructed. But in this case, we can look over the shoulder of the test developers. The test designers (Roodhardt, personal communication) first note that the construction of a good test item takes considerably more time than solving the problem. It is first necessary to find a source of potential test content. Some people have a special ability to recognize prospective sources. Libraries are good sources of material. In this particular case, the scientific magazine $H_2O$ offered an article that was used as a source. One makes this search with certain principles in mind:

- The story should fit with the philosophy of the curriculum;
- The problem must bear some relevance for the students;
- The text must inspire additional questions;
- It must be possible to design a string of good questions at examination level. Most of the time we are forced to simplify the article to the level treated in the classroom, but in this particular case, this was not necessary.

One of the people involved in the design process takes the lead in developing a first draft that illustrates the potential of the context to the other team members. After discussing the draft, it is concluded that it offers attractive possibilities. Especially attractive are the questions concerning the point at which the safety level will be passed. But a number of hurdles are still to be taken. The draft version in this case contained, for instance, the following example:

a. Each visitor delivers 0.5 g urea.
b. At the end of the day there is $1000 \cdot 0.5$ g urea.
c. Per liter water: $500 \div 1000 = 0.5$ g/m$^3$ = 0.5 mg/l.
d. Fresh water: $30 \cdot 1000 = 30,000$ l.
e. Disappearing $30,000 \cdot 0.5$ mg = 15,000 mg = 15 g.
f. Only 485 g urea is left.

—Compute in this way the amount of urea at the end of the second day and at the end of the third day.
—On which day will the statuary standard be exceeded?
—The legal standard is 2 mg/l urea. Will this norm be exceeded after three days?

It was concluded during the discussion that this was not a good format: The exemplary computation should be replaced by something more substantial. The whole problem is based on these computations. Therefore, mistakes in this phase either are not to be allowed or we must change the string of follow-up questions. The essential question is about exceeding the standard. But we could give away a little bit about the computation—give the student a checkpoint. If this fails, it tells us something significant about the level at which the student is working.

Thus, a new version begins to take shape. Among other things, the quality of water in a swimming pool is judged on the basis of the amount of urea. Urea enters the water via perspiration and urine. It appears that the average daily increase in the amount of urea is 500 g per 1,000 visitors a day. The water has to be refreshed in such a way that the statutory standard of 2 mg per liter will not be exceeded. In our model, this means that if 1,000 people visit daily a pool with a volume of 1,000 m$^3$, the statutory standard for refreshment is 30 l per person per day. This means that at night we refresh 30,000 l (or 3 percent of the total). On the first day, we begin with 0 g of urea in the water; at the end of the day, the water contains 500 g of urea. After refreshing, 485 g of urea will remain.

- Compute the amount of urea at the beginning of the first day.
- Assume that the amount of urea at the beginning of a day is $U$ g.
- How large—expressed in $U$—is the amount of urea at the beginning of the next day?
- On which day will the statutory standard be exceeded?

Upon reflection, we decided that the computation based on the standard for urea of "2 mg/l" could cause misunderstanding that, in turn, might cause problems when grading the tests. So this was changed to "2 mg/m$^3$."

In the final discussion, the test takes the form described earlier. The first two questions are partially answered by telling the students to show that at the beginning of the third day the amount is 955 g. And they are asked explicitly to explain their answers. Also, the decision was made to use a photograph with the test, in the hope that this would have some psychological effect on the students.

One aspect of this item that was discussed in detail was context. It was clear that context plays a relevant role and that the students would recognize the real-world quality of the problem. Its context was both real and relevant.

*Real versus Artificial*

It is clear from the previous discussion the extent to which the authors considered context in the design of this problem—the relevance and the reality of this problem for the students for whom the examination was designed. In general, the problem was well received both by students and their teachers. But the context of the following problem was a different matter!

- As the result of the increase in space travel in the twenty-first century, a new disease from outer space struck the inhabitants of the earth. The graph shows the number of victims of this disease on the planet for the years 2079–2086.

  —Draw a new graph on logarithmic paper for the number of victims.
  —During which period is the increase in the number of victims nearly exponential?
  —Compute up to one decimal point the annual growth factor in cases during this period.

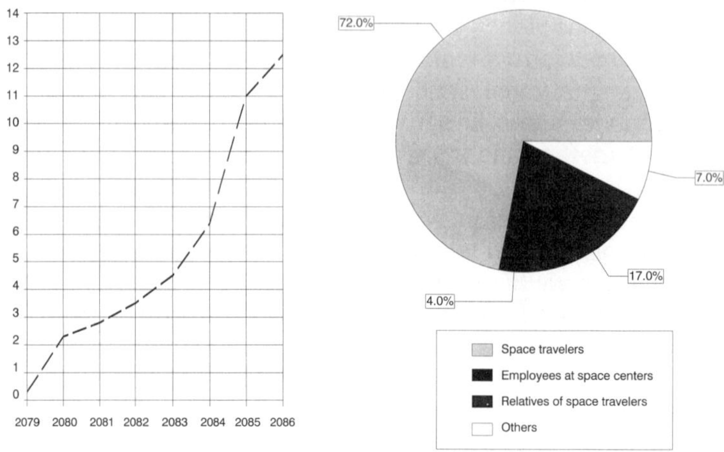



Those suffering from the disease were primarily space travelers and employees at space centers. The pie chart shows the distribution of victims in 2086. The number of infected people in The Netherlands in 2086 was space travelers, 60; employees at the space center, 5; relatives of space travelers, 3; and others, 2.

—Make a pie chart for the situation in The Netherlands.
—Investigate whether in 2086, among patients in The Netherlands, there were significantly more space travelers than the 72 percent for the whole earth, with a significancy of 1 percent.

In a hospital, the disease is treated with the medicine R1 and R2. Every patient gets 600 mg of R1 and 190 mg of R2. Both medicines can be made from raw materials A and B. Every kg of A yields 60 mg of R1 and 10 mg of R2 and every kg of B yields 30 mg of R1 and 15 mg of R2. Compute the minimal number of kg of raw material (A and B together) needed for one patient.

The cost for A is $15 per kg and the costs for B are variable. The hospital tries to get the raw material for minimal cost per patient. Compute for B when it is cheaper to make the medicines R1 and R2 from A only. (VWO National Examination, The Netherlands, 1987)

The context is clear from the first sentences, but we have presented the complete problem to give an honest picture of the process. A first reaction to the context might be that it is artificial to project a century from now, to talk about a space-related disease. It might even appear to be an elaborated problem because of contrived information, such as R1 and R2 and A and B. It definitely does not offer the students a real-world context—apart from the fact that the mathematics is all too real for them. And the argument can be made that its relevance leaves something to be desired.

Nevertheless, the information presented in the problem is scientifically grounded and based on a source article the contents of which are both very real and relevant. The problem was built around real data: The time span was 1979 to 1986, and the space disease was in reality AIDS. But the designers of this item felt that it was not a good idea to confront students under examination conditions with this highly emotional issue. And here we reach the heart of the matter.

It seems evident that when we put so much emphasis on the importance of mathematics education in preparing our students to be intelligent and informed citizens, we have to deal with all sorts of real-world contexts. We have to deal with pollution and its very political implications. We cannot avoid politics, with its multitude of subjective components. Traffic safety is an important matter in general, but one with a very emotional component for the many students aware of casualties in their families. Health is perhaps one of the most important issues at this time for many people. The fitness trend is still very strong and identified with positive bodily care. But to discuss cancer, Alzheimer's, heart disease, and for that matter the effectiveness of certain treatments—whether or not in relation to the costs of health care as a political issue—presents subtle and not so subtle challenges. We recall vividly an incident that took place in one of our experimental schools. Statistical data presented in a textbook problem showed an exponential growth in the number of abortions in different countries (excluding the students' own country) and those numbers (not the subject of abortion) were discussed. The page on which this problem occurred was torn out of the book at one school because a student or student's family believed that abortion should not be discussed at school and certainly not during mathematics.

Over time, we have noticed a gradual change in that attitude: A greater and greater number of real issues can be discussed at school, if we remain attentive to their emotional, psychological, and political aspects. However, it is also clear that at the teaching and learning level, we may be able to use contexts that are not possible at the assessment level. We agree, for instance, that in 1987, to use AIDS as a context was not without risks because of the highly emotional and uncertain aspects of the disease at that time. Now, only five years later and with somewhat more knowledge regarding the growth or nongrowth of the disease, we can imagine the possibility of AIDS as a context in a classroom discussion or maybe even on an examination. But it is clear that there is risk in using real contexts on tests.

Another test item that further illustrates the kinds of problems we face in bringing the world into the classroom does not seem to deal with a volatile context.

- In a certain country, the national defense budget is $30 million for 1980. The total budget for that year is $500 million. The following year the defense budget is $35 million, while the total budget is $605 million. Inflation during the period covered by the two budgets amounted to 10 percent.

  —You are invited to give a lecture for a pacifist society. You want to explain that the defense budget has been decreasing during this year. Explain how to do this.
  —You are invited to lecture to a military academy. You want to explain that the defense budget has been increasing this year. Explain how to do this. (de Lange, 1987; see also MSEB, 1991. Used with permission.)

This item precipitated a number of conflicts in the classroom originating from the basic question: Is it ethical to teach students how to be manipulative themselves rather than to show them examples of "manipulation" by others? In the process of teaching them to recognize how data may be manipulated, is it appropriate to ask them to do it themselves? These questions are directly related to the problem of real contexts: Should we discuss such a controversial matter as defense spending in a classroom situation, let alone use it in assessment? The following student discussion bears thoughtful scrutiny:

> *Marijn*: I think you have to see it as a percentage—30 of the 500 and 35 of the 605.
> *Marc*: 500 of the 30, that's 100/6.
> *Marijn*: The other way around—that's 0.06.
> (Marc then calculates 35/605 on his calculator: 5.78.)
> *Marijn* (says to Susan): Write that down.
> *Susan* (asks): What is that the answer to?

(Marijn tells her and then dictates the answer.)
*Servaas:* This one is really too simple.
*Marijn:* Aren't we supposed to do something with the
inflation?
*Marc:* O, (expletive)!
*Servaas:* If you ask me, that has nothing to do with it.
*Marc:* The inflation applies to both amounts so they
cancel each other out.
*Servaas:* In the second one, it just increased from 30 to
35.
(Susan doesn't agree at all: The inflation lies in be-
tween the two numbers, so you have to figure it out
for the second one.)
*Marc:* And you have to add 10% extra to that 605.
*Servaas:* It doesn't say that.
*Marijn:* But in the next one you have to do it.
*Marc:* You add on the inflation, but you don't mention
that there was any inflation, so the difference is even
larger.

The arguments held by the pacifist group and by the military
academy overlap each other somewhat, making the role to be
played by inflation rather unclear. The following four minutes
hardly contribute to the solution. Marijn calculates 605 and 35
backwards on a basic annual level (605, that's 100%, so you divide
that by 11 and then subtract it, so that used to be 550) and then
establishes that 31.8/550 is 5.78. But Marc had already pointed this
out in the beginning with his "canceling out." Marc does, however,
get an idea from Marijn's calculations:

*Marc* (says to Susan): If you subtract the inflation from
35 you get 31.8. This 31.8 is much less in relation to
605 than 35. So you have to subtract inflation from the
35 but not from the 605.
*Susan:* That sure is stupid.
*Marc:* Yeah, but you have to do your best to sell it, so it
should be O.K. to fiddle a bit.
*Susan:* That's ridiculous.
*Servaas:* You can't do that.
(Marc leafs through the booklet: They are doing that
all the time.)
Marc: O.K. It's not all right, but if you're on the side of
the pacifists. . . .
*Servaas:* But then calculating the inflation for both.

> (Marc asks Marijn what she thinks, but she had lost the thread of the conversation. He explains it once more but Marijn, too, rejects his solution: That's no longer objective.)
> *Marc:* But they don't have the data.
> *Servaas:* And at a certain point you say whatever seems to fit.
> *Susan:* I'll write it down.
> *Marijn:* Marc sure knows how to deal with pacifists.
> *Susan:* Double points. Go ahead Marc.
> (Marc dictates his solution.)

The discussion shows that the context is all too real for the students but they hesitate to use the numbers to their benefit, making the assumption that you have to be objective. This conversation was taken from a classroom where the students were working on solving this problem in small groups. One can imagine what would happen if we had given this problem as part of our assessment, whether as group work or individually.

Let us turn finally to a problem within an artificial context:

- Somewhere out in a remote area, where people are rarely seen, a mysterious factory exists. Above the entrance hangs the sign "Cote d'Or." According to whispered rumors, the alchemist Ben Al-K'wasi is creating golden Christmas ornaments out of clay by means of a complicated procedure. The new ornaments are made entirely of clay. After a year of maturing, they turn silver and, after one more year, they become true gold! If they do not break, they will remain gold. If an ornament breaks, it dissolves forthwith. In 1983, the factory attic contained 217 clay ornaments, 128 silver ones, and 70 gold. By means of extremely uncommon earthly rays, the silver and gold ornaments can produce young: Two silver ornaments are able to produce one young clay ornament. In 1984, there were 288 clay ornaments and 98 gold ones. In 1985, there were 230 silver ornaments. Due to various causes, 70 percent of the golden ornaments break yearly.

    —Based on this data, draw up the corresponding Leslie matrix (for a period of a year).

—Calculate the missing data for 1984 and 1985 (From a school examination in The Netherlands, 1986).

This problem may be successful in an appropriate classroom climate, where such exercises are part of the didactical contract between the teacher and the students. It is, however, an almost perfect example of an artificial context, which may irritate students. In addition, with nothing to support the context, there is certainly no need to reflect. The imaginary context would tend to distract rather than support their efforts, and the teacher would have trouble analyzing the results.

### The Distance to the Students' World

If we simply reflect on the examples provided, it will be clear that a valid topic for discussion is, How real is the real world for the student? The American *Curriculum and Evaluation Standards for School Mathematics* (NCTM, 1989) stresses the importance of motivating contexts like pop music charts and baseball. A motivational factor may be necessary when we use a problem to introduce mathematical concepts, but when it comes to applications we cannot be too prudent in describing the boundaries of the students' real world.

We mentioned previously that flying formed a real-world context that is very rich in relation to mathematics. Late in the 1970s, experiments were carried out putting simple trigonometry and vectors in the context of flying, aiming at lower and middle ability students of 13–14 years of age. The experiments were satisfying and a booklet on them was published. Shortly thereafter we received complaints to the effect that putting mathematics in a flying context was not fair because this subject puts boys at greater advantage and was unfit for girls. Although the message was clearly understood, it came as a complete surprise to those who complained that the booklet was tested at a school where the students were almost exclusively girls.

But although the dilemma was clear, it still was not satisfactorily resolved. Should we offer girls only female contexts in order to emancipate them? Or is it much better to offer them a supposedly male context for emancipation. During the experiments, we found that the context motivated boys and girls equally and that boys enjoyed no major advantage because of the way the information was presented.

Since that time, we have solved the problem (for the moment) by offering as many different contexts as possible—both during the classroom presentations as well as during assessment. Slowly, rather specific "female" contexts are entering the tests—especially at the primary level, where this problem seems easier to solve than at higher levels. This may be caused in part by the fact that at primary level, and at lower secondary level as well, we tend to define the real world as the world that is really known to the students, or can be imagined by them. In other words, the distance to the students' world is close to zero.

Several test items on the primary level that have been selected at random follow:

- An ice cream vendor has computed that if he sells 10 ice creams, they will consist of: 2 cups, 3 cones, and 5 sticks. He orders 700 ice creams. What distribution of the different kinds will he use?

- You need to know how much water your water barrel can contain. What are you going to compute?
  —the perimeter
  —the area
  —the volume
  —the weight

- Annie wants to know the area of the island. She does that by putting a grid on the map of the island. What will the area of this island be?

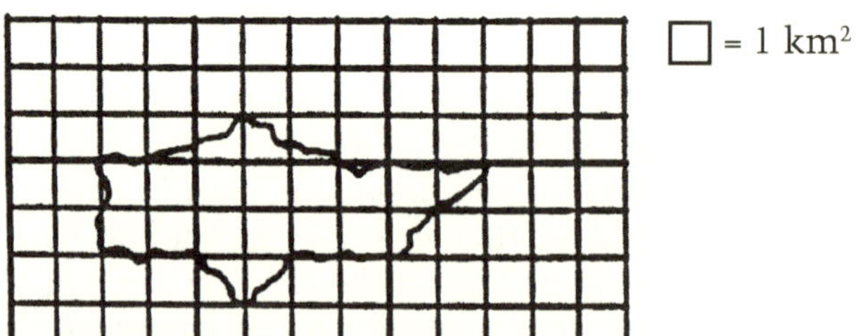

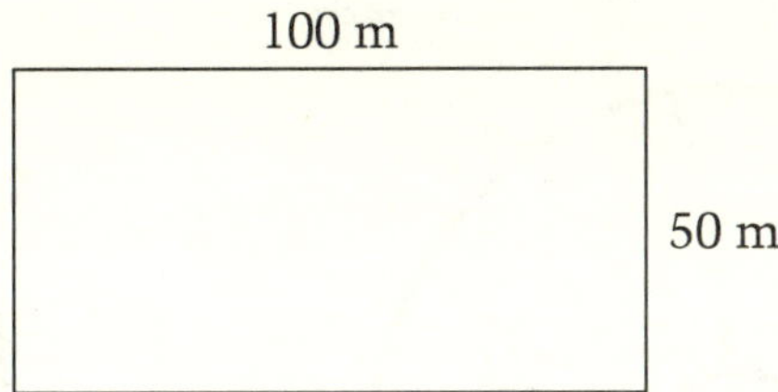



- Frank runs around this sports field five times. How many kms did he run?
- A pack of papers containing 500 sheets is 5 cm thick. How thick is one sheet of paper?

All of these problems are situated, more or less, in the students' daily life. But the quality or nature of real contexts changes as we advance in the educational system. We have already seen the different real worlds that students have to cope with if they are at the high school level. After observing these developments rather carefully in The Netherlands during the past decade, we have seen the following picture evolve: At the primary level, students are dealing with their "own" real world, including fantasy worlds. But at secondary level, the picture is different: In the first place, we notice that the students are becoming increasingly a part of the real world, including the scientific and political worlds. But we also see that, in effect, we delay this process for the lower ability groups. Here we stay much longer with the day-to-day real world, without any assurance that this is justified. At the same time, we are seeing female real worlds appear in tests for the first time. The following is taken from the new final examination at the lower level in The Netherlands in 1991.

- Wilma's sister has joined the majorettes and therefore needs a circle skirt. Wilma has promised to make her one. She has made a sketch of the pattern of the skirt and has indicated the measurements: waist 56 cm, length of the skirt 40 cm.

  —Draw the pattern for the half circle of fabric to scale. Write down your calculations.
  —Wilma buys a piece of cloth measuring 90 cm width. How long should the piece be? (round off to 10 cm) (W12–16, team, 1991. Used with permission.)

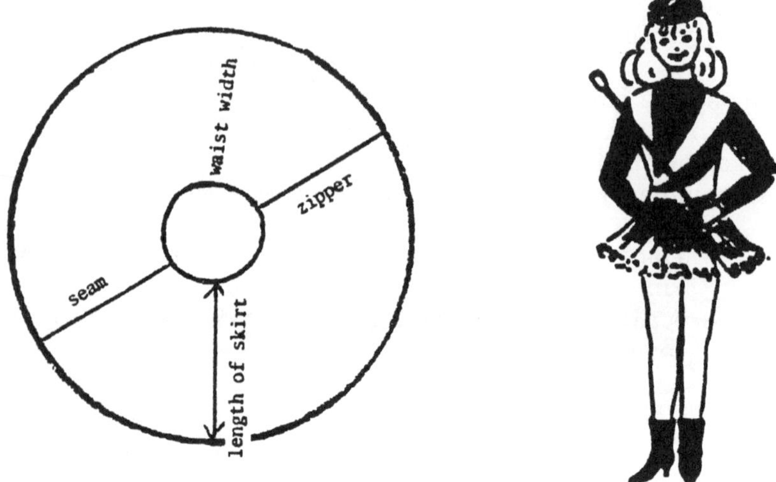


This context, from daily life, is oriented to girls—close to their real world; it is not as relevant to boys. The next item, meant for a little lower level, illustrates already a somewhat more "scientific" world:

■ The flowers on this stem grow next to each other:

—The first flower uses $1/3$ of the nutrition that is transported through the stem. The second flower uses $1/3$ of the nutrition that is left over. The third flower uses $1/3$ of what is left over by the second, etc.

—Complete the table:

| Flower # | 1 | 2 | 3 | 4 | 5 | 6 |
|---|---|---|---|---|---|---|
| Part of Nutrition | 1/3 | 2/9 | | | | 32/729 |

(Boertien, National Center for Educational Evaluation, CITO, Arnhem, The Netherlands, 1990. Used with permission.)

It will also be clear that not only is the context somewhat less close to the student, but that the context is rather artificial and that nothing relevant is done with the context. A similar context—the growth of plants—is used in a very different way in a standardized examination, where it is clear that, although the proximity to the students' world is remote, the problem may nevertheless be very real to them.

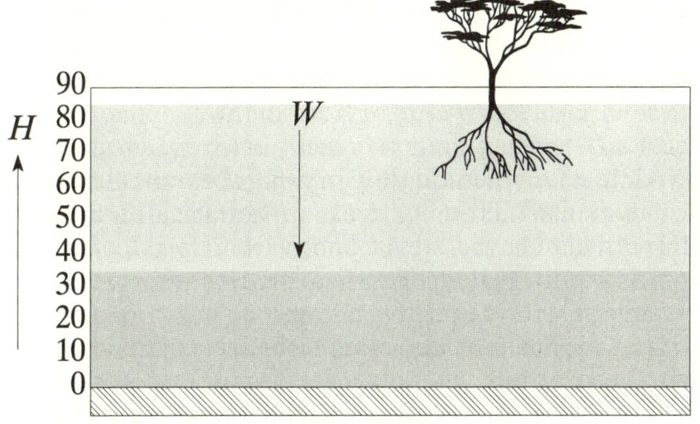


- In a nature area, the groundwater level lies at a soil depth of 90 cm. Ten cm above the groundwater level, the moisture content of the earth is approximately 32%. The higher the ground above groundwater level, the lower is the moisture content of the earth. At 80 cm above groundwater level, for instance, the moisture content of the earth has decreased to 4%. The relation between the height of the groundwater level and the degree of moisture content is indicated by the formula:

$$H \cdot p = 320$$

Here $H$ indicates the height above the groundwater level, expressed in cm, and $p$ indicates the degree of moisture content, expressed in percentages. The formula can be used for $H$ between 10 and 80. Draw a graph of the relation between $H$ and $p$ on the figure on the worksheet.

The area is going to be planted with vegetation whose roots require a moisture content between 5 percent and 10 percent at their maximum depth. Calculate which heights above groundwater level will be suitable for this.

The maximum root depth (in cm) of a plant in this nature area we will call $R$. Give a formula where $p$ is expressed in $R$.

The groundwater level in this nature area is now raised 30 cm. The relation between the height above groundwater level and the degree of moisture content remains the same as in the earlier situation. Calculate the new degree of moisture content of the earth at a depth of 40 cm (HAVO National Examination, The Netherlands, 1990).

The "distance" of the real world is a factor that we have to consider, together with the degree of "reality" of the context. It is very difficult to be sure when a context is real and when it is close to the student. We must bear in mind that, in general, we should try to use real-world contexts, but the AIDS example made clear how complicated this can be. On the other hand, what is real for one student is not necessarily real for another. This reinforces the importance of offering a wide variety of contexts—hardly a revolutionary conclusion, but one that in fact is difficult to implement under assessment conditions.

## NECESSARY AND SUFFICIENT

If we look at the previous examples, we find very few that do not contain all of the necessary and sufficient information. It is so natural to assume that we need all the information in the exercise and that all information we need really is out there that we hardly think it worthwhile to consider problems that are not of that form. Even in more complex real-world problems, we can solve all problems by analyzing carefully all the information, mathematizing and organizing it if necessary, and using specific mathematical tools and techniques to solve the problem. If we reflect for a moment on the problem of lecturing to the pacifist society and military academy, we notice that after a suggestion for a quick and clean solution, the students were unsure because they did not use the information regarding inflation. They were convinced that something had to be wrong if they did not use all of the available information.

If we teach real problem solving or, better still, if for the most part we identify mathematics with problem solving, we have to bear in mind that usually the solutions do not come easily. In real life, we have to mathematize the problem and that means in the first place analyzing it to identify the relevant mathematics. This is a very difficult task and almost neglected in mathematics lessons.

One issue is that of structuring a problem. In one or two previous examples, we have seen a structure that was provided to help the students get started, but it is apparent that problems can be rendered more complex and real by omitting this prestructuring. This touches immediately the essence of the paradox involving testing and real-world problems: How do we present problems to the student in such a way as to optimize his or her chance of successfully solving them? This is a hard question to answer in the standard teaching-learning situation, but even more so in a test situation. It is more or less widely accepted that we should offer the students the possibility to at least start successfully. To illustrate this point, it suffices to analyze here some of the items presented that seem more complex; it will become clear how carefully the designers selected the initial questions. These first questions function more as confidence builders for students than anything else. A further illustration of this point is the earlier discussion on urea pollution of a swimming pool.

Thus, a question of major importance is, How can we offer students good tests that contain problems that are more or less real, as well as some guarantee that the students can make sensible efforts at solving them? An example that I believe illustrates such a problem is the following snowplow problem from the Mathematical Contest in Modeling:

■ The solid lines of the map represent paved, two-lane county roads in a snow removal district in Wicomico County, Maryland. The broken lines are state highways. After a snowfall, two plow-trucks are dispatched from a garage that is about 4 miles west of each of the two points (*) marked on the map. Find an efficient way to use two trucks to sweep snow from the county roads. The trucks may use the state highways to access the county roads.

Assume that the trucks neither break down nor get stuck and that the road intersections require no special plowing techniques (Chernak, Kustiner, & Phillips, 1990. Used with permission.)

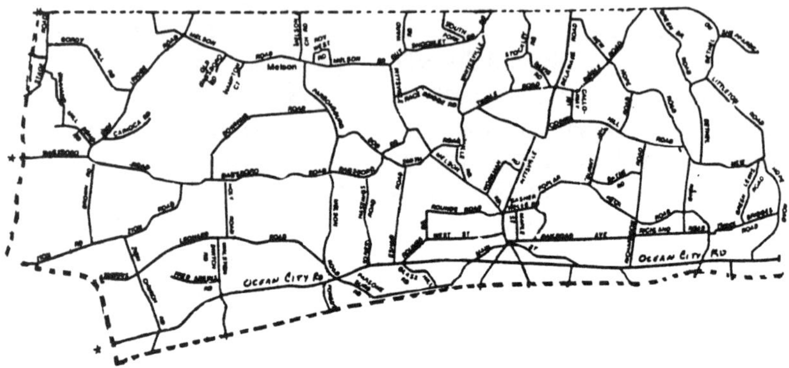

It will be immediately clear that a lot of information is lacking if we really want to solve this problem in a realistic way. Two assumptions are already fixed: there are no breakdowns and no special techniques at intersections. But if we look at the results that college undergraduates turned in, we notice that a lot of information is lacking that is at least as important as that provided. We note that

- —there are two state highways
- —the state highways are clear of snow when the plows are at work
- —all county roads are paved, two-lane, two-way
- —no new snowfall occurs after the plows begin
- —there is clear weather (no accidents or interference)
- —all roads need to be plowed
- —the county is flat
- —there is no mountainous area
- —the plows may turn right or left, and may turn around at intersections
- —each truck has a 60-gallon fuel tank and averages 3 mpg
- —each truck is equipped with a plow blade set at an angle of 45°

All of these assumptions were made by one team.

Tests or tasks with redundant information are even harder to find. This may seem strange because, in real life, we usually have to solve problems with lots of redundant information. And some evidence exists that it is not only students who find problems with redundant information difficult to solve. One example, described in de Lange, Burrill, Romberg, & van Reeuwijk (1993), is called the *rat problem*. (Used with permission.)

During inservice teacher training courses in the early 1980s, the following from a college textbook on biology was given upper secondary mathematics teachers:

- [I]t might be interesting to estimate the number of offspring produced by one pair of rats under ideal conditions. The average number of young produced at a birth is six; three out of those six are females. The period of gestation is 21 days; lactation also lasts 21 days. However, a female may already conceive again during lactation; she may even conceive again on the very day she has dropped her young. To simplify matters, let the number of days between one litter and the next be 40. If, then, the female drops six young on the first day of January, she will be able to produce another six 40 days later. The females from the first litter will be able to produce offspring themselves after a 120 days. Assuming there will always be three females in every litter of six, the total number of rats will be 1,808 rats by the next January 1st, including the original pair. . . .

  Is the conclusion that there will be 1,808 rats at the end of the year correct?

During the teacher training course, only 20 percent of the teachers were able to solve this problem within half an hour. As they explained: "We feel we have all the tools to solve the problem, but we are unable to use them."

On the other hand, nonmathematics majors (16 years old) who had had about a year of "new" real-world mathematics education at some of our experimental schools did very well on the problem. Results depend, of course, on the test conditions. In the classroom, with a limited amount of time, students find it very difficult to solve, or even to schematize the problem, but with no time limit (for instance, having the problem as homework or a take-home test), students produce fine results. This indicates that such process-oriented activities are not well suited for testing by means of time-restricted written tests.

One girl came up with the solution in figure 4.2, which is surprisingly simple.

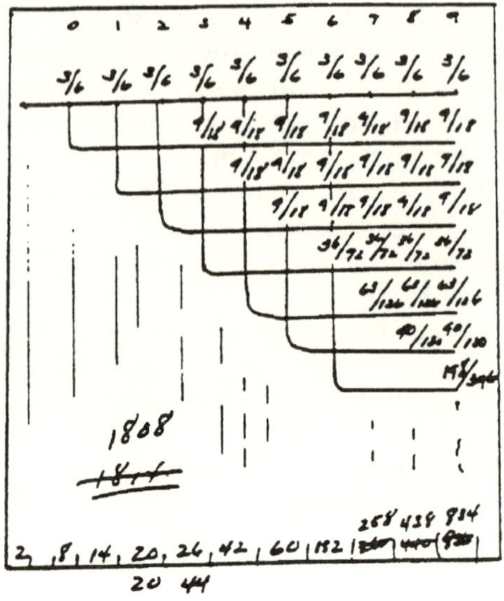

Figure 4.2  One solution to the rat problem.

The schematized solution in figure 4.3 is from a teacher. The teachers felt the need to produce a formula and consequently came up with:

$$A_{n+3} = A_{n+2} + 3A_n = 2$$
$$A_0 = 8$$
$$A_1 = 14$$

| $t$ | -1 | 0 | 1 | 2 | 3 | 4 | 5 | 6 | 7 | 8 | 9 |
|---|---|---|---|---|---|---|---|---|---|---|---|
| N | 2 | 6 | 6 | 6 | 24 | 44 | 60 | 132 | 258 | 438 | 834 |
| T | 2 | 8 | 14 | 20 | 44 | 86 | 146 | 278 | 536 | 974 | 1808 |

Figure 4.3  Another solution.

A completely different approach uses graphs and matrices (which are part of the "new" curriculum). The following graph represents the growth of the rat population and the graph can be represented by a matrix:

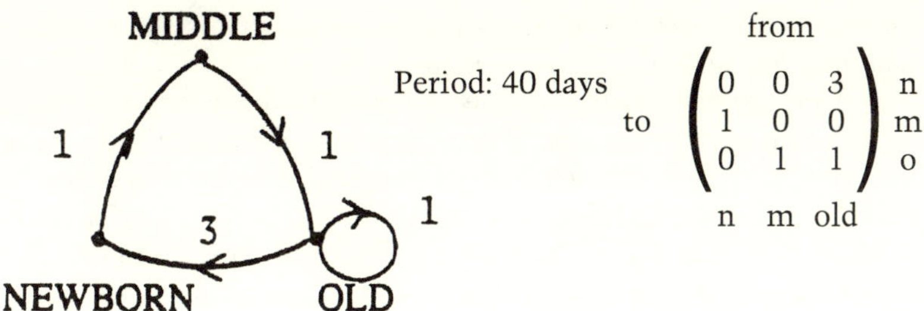


$$\begin{pmatrix} 0 & 0 & 3 \\ 1 & 0 & 0 \\ 0 & 1 & 1 \end{pmatrix}$$

Another possibility is to look for the nature of the growth process. Comparing the number of rats period by period, we find that the growth factor in the long run is equal to 1.86. This leads to the formula

$$A_n = 44 \cdot 1.86^{n-3}$$

We leave it to the reader to integrate and generalize the different solutions—an activity that is representative of higher level mathematization.

The biggest problem for most teachers was to mathematize the problem in the first place. Here was a story with mathematical aspects, but it confronted them with the problems: Which part is relevant, and in which way are the relevant parts connected? They also were very frustrated because they were unable to use the powerful mathematical tools at their command to solve this problem and were reluctant, at least initially, to accept the girl's solution as a proper and even beautiful solution. Later, teachers offered us very elegant solutions integrating all kinds of mathematics: matrices, graphs, Pascal's triangle, characteristic equations, eigenvalues, and much more. This opened up another area of concern—how to compare and value such different solutions. Many teachers were inclined to consider "mathematical" solutions superior to solutions that did not use typical "mathematical tools," like the girl's solution. This clearly illustrates that the belief systems of teachers are harder to change than those of students—not to mention those of test designers.

Everyone seems to agree that we should have more "rat problems," and most certainly during the teaching and learning process. But even then we face the matter of comparing the results. Is the girl's solution really "without" mathematics, as some teachers state? Or, given its simplicity, is it actually one of the best solutions, as some other teachers argue? It is not at all clear, at least on the "proof" level, that the solution of the girl is the result of teaching her realistic mathematics.

Here another issue emerges. How can you prove that your educational efforts have changed the problem-solving attitude of the students? Although we have very strong feelings about that aspect of realistic mathematics education and mounting evidence of its effectiveness, it seems also clear that it is almost impossible to produce hard evidence of its efficacy. Of course, one has only to look at the central examinations in The Netherlands to see that they have changed considerably and that they are in fact testing problem-solving abilities. But such tests are still far from tests like the rat problem.

The teachers in our training course questioned where the girl had learned to visualize the rat problem the way she did. Undoubtedly, she had never encountered a similar problem, but she was used to flexible use of representations and was able to transfer those capabilities in quite another setting. The point to be noted here is that the girl was not the only one with an elegant solution to the problem: The majority of the students in her class solved the problem with some kind of schema or visualization. But several had managed it with simple, plain language and minimal mathematical notation.

It seems clear, perhaps disappointingly clear, that we are only at the beginning of an exploration of this aspect of mathematics education and even further away from its consequences for assessment.

An interesting example of a lack of fit between the information provided on a test item and the information needed for its solution is illustrated in a problem for grade 6:

- Katie bought 40 cents worth of nuts. June bought 8 oz. of nuts. Who bought more nuts?

  a. June.
  b. They each bought the same amount of nuts.
  c. Katie bought twice as much.
  d. Katie bought 5 oz. more of nuts.
  e. You cannot tell which girl bought more nuts (Adapted from 1989 Illinois State Board of Education testing materials, with permission.)

This is interesting from different perspectives. In the first place, it is encouraging that an American state board of education had the temerity to try such an item. It is a breakthrough to give a problem lacking sufficient information, certainly for grade 6. But, of course, it raises certain questions, too. In the first place, what does it tell the teacher that 61 percent of the students have answered correctly: e? Or more precisely, what are we measuring here and how certain are we that the correct answer reflects the reasoning level being tested?

The idea behind the item is certainly appealing, but the multiple-choice format destroys its effectiveness. Imagine that the item had been expanded as follows: "Each of the following four answers are correct if certain assumptions are made. Describe in each of these four cases the necessary assumptions."

Now we have a completely different item. The students have to reason, to think, to write down their reasoning. With just a slight alteration, we have created a test item that operationalizes higher order thinking skills as well as communication skills. Of course, we lose ease and efficiency in grading. And we might even lose some objectivity in grading. But we gain so much more. Look at some of the possible answers:

> June bought more nuts if the nuts are more expensive than 40 cents for 8 oz.

> June and Katie bought the same amount of nuts if the nuts are 40 cents per 8 oz.

> Katie bought twice as much as June if the nuts are 20 cents per 8 oz.

> Katie bought 5 oz. more of nuts. So she bought 13 oz for 40 cents.

Modified as suggested, this simple test item would challenge the student to process the information and communicate these processes to others. And the teacher, rewarded with real feedback on the level of the student, subsequently can inform parents of their children's progress in a more informed way than with a score from a meaningless multiple-choice item.

## FORMATS OF TESTS

### Multiple Choice

In constructing an achievement test to fit a desired goal, the test maker has a variety of item types from which to choose. Multiple-choice, true-false, and matching items all belong to the same category—selection-type items. Officially, they have become popular because they can be scored objectively. This means that equally competent scorers can score them independently and obtain the same results. These equally competent scorers are usually some machine. And therein lies the real popularity of selection-type items: They can be machine scored and therefore are very cheap to administer.

The rules for constructing a multiple-choice item are simple. A multiple-choice item will present students with a task that is both important and clearly understood, and one that can be answered correctly only by those who have achieved the desired learning (Gronlund, 1968). That this is not as simple as it seems to be, we all know, especially if we hold that the item should operationalize a specific goal.

To show how difficult the latter seems to be, we give an example based on an item in the second IEA study (Travers & Westbury, 1989).

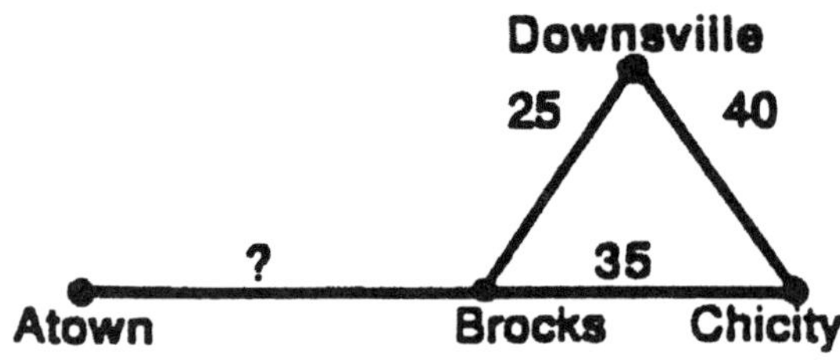


- John and Mary make a trip by car. They go from Atown to Brocks, then to Chicity to Downsville, and back to Atown. The total trip is one of 190 miles (Board of Trustees, University of Illinois, 1989. Used with permission.)

—What is the distance from Atown to Brocks?

  a. 35
  b. 40
  c. 45
  d. 55
  e. 70

Of course, the item would be a winner if it were not for the multiple-choice format. But what is far more serious is that this item is meant to operationalize the goal: linear equations. So the test-item designer is not only the test designer but also the solution designer. Many educators would prefer that students solve this problem in their own way, which, in this case, would seldom include linear equations. More serious is that the item, as illustrated, is flawed—seriously flawed, because we no longer know what is being measured (which is not always bad), but we pretend that we do know (which is very bad). "American students are very poor in linear equations" could be a meaningless statement if it were based on items like this.

Another frequently mentioned problem with multiple-choice items is the assumption that only those who have achieved the desired learning can answer the question correctly. A Dutch teacher collected examples of student reasoning behind the answers they gave on a nationwide standardized multiple-choice test (Querelle, 1984. Used with permission.)

$\{0\}$ is the solution of

a. $3x = 3x$
b. $-3x = 3x$
c. $3x = 3x + 1$
d. $-3x = 3x + 1$

Most students chose a. But, fortunately, some students did choose b. Rudy, certainly not one of the brighter students, was asked to explain his choice of b:

Why b?

Well, because that's the only one that's 0.

Yes, but how come you're so sure?

Well, that's easy. With a, you get $3x + 3x = 0$, that's wrong. And with b, you get $-3x + 3x = 0$, so that's the one, because the others are wrong too.

Let us look at a final example:

■ The radius of a cylinder is 1 cm. The height is 2 cm. The cover of the cylinder is a rectangle. What is the length of this rectangle?

　a. $2\pi$ cm.
　b. $4\pi$ cm.
　c. $\pi^2$ cm.
　d. $2\pi^2$ cm.

*John:* I have a as the answer.

Okay, let's hear it.

I did the area, so radius multiplied by radius multiplied by $\pi$, and, well . . . uhh one times one is two, so two $\pi$. So a is correct.

*Babette:* I have got a as an answer too, but different again. I thought also area, so $1 \times 1 \times \pi$ and that doubled for the height.

The construction of a multiple-choice item that is clearly understood—answered correctly by those who have achieved the desired learning—and operationalizes a specific goal or learning outcome is not a simple task. Many items have flaws, and all have very limited value if our objective really is authentic assessment. At this time, the only area of application seems to be operationalizing the lower goals. And Travers and Westbury (1989) state, when discussing the second IEA study, "The construction and selection of items was not difficult for the lower levels of cognitive behavior-computation and comprehension." (This is not completely in line with the items as we see them; we refer once more to the linear equations item.) "But," they continue, "the difficulties were presented at the higher levels. The multiple-choice format necessitated by the scale of the study presented some opportunities as well as challenges." They then give an example:

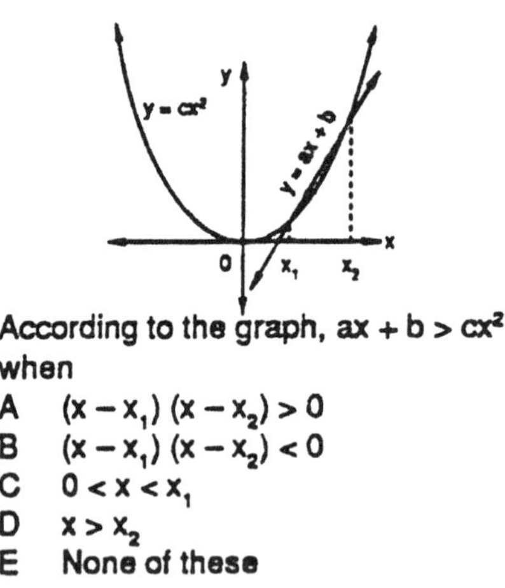


According to the graph, $ax + b > cx^2$ when

A    $(x - x_1)(x - x_2) > 0$
B    $(x - x_1)(x - x_2) < 0$
C    $0 < x < x_1$
D    $x > x_2$
E    None of these

According to Oldham, Russell, Weinzweig, and Garden (1989), this item as an open-ended problem would involve reading the graph;

that is, it would require comprehension behavior. If $x_1 < x < x_2$ were included as one of the choices, the item would involve the computation or comprehension level of behavior. However, given the choices offered as solutions, one has to analyze them and discover the relationship they determine. This involves behavior on the analysis level. In this case, the graphs of the functions become one way of communicating certain information that must be matched to other descriptions of the same data. This item provides an example of the imaginative use of a multiple-choice format to yield a problem at a higher level of behavior than an open format would have used. This statement, however, needs further analysis, just like the item itself. Let us first look at it from the students' point of view. First, it is evident that C and D are not correct; if C and D had been combined in one alternative, there might be reason for giving such a response some consideration. What we see at once—and it is a trivial matter—is that the answer should be $x_1 < x < x_2$, if we understand the notation involved. How does this relate to A and B? Easy: Take 1 and 3 for $x$ and see what happens in between if you take 2 for $x$. For B, this yields $(2 - 1)(2 - 3) < 0$. Done.

Now the question is, What "behavior" has been tested? Did we indeed soar to the analysis level? Of course not! We were at the lowest level, substitution. Did we need to match certain information to other descriptions of the same data? Not at all. Was it an important question for the students? No, because we were not dealing with a real problem. Was is a good item? Of course not. Did it operationalize the intended (high) goal? Definitely not.

In conclusion, for us the case is not complicated. There is a place for the multiple-choice format. But only in very limited contexts and only for measuring the lowest learning outcomes. For higher order goals and more complex processes, we need other test formats—the most simple being "open questions."

### (Closed) Open Questions

Multiple-choice items are often characterized as closed questions. This suggests that there are open questions as well. However, we have to be careful with such terminology: Sometimes the question is open in format, but closed in nature. A pair of examples will serve to illustrate this point:

- a. How many diagonals does a rectangle have?
  b. How many diagonals does a square have?

- The point (2, 3) lies in the first quadrant (I). Indicate for each of the following points which quadrant they occur in.

  a. (–2, 3)
  b. (2 ,–4)
  c. (3, 4)
  d. (–5, –6)

These are rather extreme examples of so-called short-answer questions, a subdivision of open questions. Although they are, technically speaking, open questions, they are very closed by nature. The respondent has to answer by a number, a yes or no, a definition, or maybe a simple graph or formula. Hardly any thinking or reflection is involved.

The distinction between (closed) open questions and (open) open questions is admittedly rather arbitrary, but this does not mean that we should not pay attention to it when designing tests.

### (Open) Open Questions

In our view, an (open) open question differs from the (closed) open question with respect to the activities involved in obtaining an answer. This answer can still be just a number or formula, but the process of obtaining it is slightly more complicated or involves higher order activities. This category differs from the next—that is, extended-response open questions—in that in the latter category, we expect the students to explain their reasoning process as part of their answer. To illustrate the difference, we submit the following examples:

- A fruit vendor buys a box of 6 pineapples for 60 cents. $x$ pineapples are defective and therefore unsalable, but the rest he sells at 18 cents each. Write a formula for his profit, $P$ cents.

- A 400 m running track is to have two parallel straights of 80 m each and two semicircular ends. What should be the radius of the semicircles?

- The length of the equator is about 40,000 km.

  —How many kilometers is it from the North Pole to the equator, measured at the surface of the globe?
  —Imagine you can travel right through the earth. How far is it, in that case, from the North Pole to the equator? (MAVO, National Examination, The Netherlands, 1991).

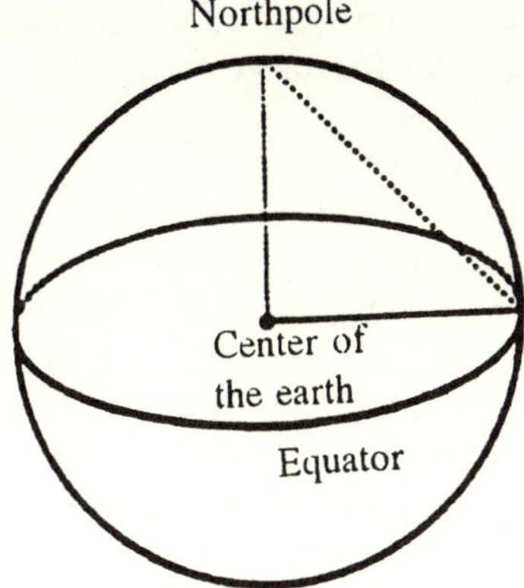



The differences between the examples given in this category and the previous (closed) category are clear. The (closed) open questions referred to basic facts and knowledge: a definition, a simple drawing, a substitution involving neither process nor thinking. The (open) open questions, although also requiring a short answer, were not just triggering questions, but questions that needed some thought and understanding and offered some possibilities for the student to solve the problem in his or her own way.

Some of the questions on the National Examination involve not only some reasoning, but the students have to explain their reasoning. This "writing down" part can be regarded as a separate goal in mathematics education, and it is a very valid one. If we stick to short-answer questions, we are not able to operationalize the communication goals. Extended-response open questions offer that possibility.

*Extended-Response Open Questions*

We offer an example from a national examination that represents the extended-response category, but with even less freedom than the fish farmers problem on page 106–107. It shows clearly how many different test formats we actually have at our disposal.

> In fall, the grapes that are ripe are harvested. The taste of the grapes depends to a great extent on the moment that they are harvested. If they can be left out in the sun a little longer, the taste will improve. But if one waits too long with the grape harvest, there is a chance of damage caused

by heavy and lengthy rainfall. A grape farmer has the following choices for harvesting:

I. Immediate harvest—
The quality is "reasonable." Half of the harvest can be sold for direct consumption; the proceeds in this case are $2 per kilo. The other half of the harvest can be used only for grape juice; the proceeds for this part are $1.30 per kilo. With these options, there is no risk involved in the harvesting.

II. Wait two weeks before harvesting—
The quality of the grapes in this case is "good." The complete harvest can be sold for $2.30 per kilo. But to wait two weeks involves some risk. If it rains more than two days during these fourteen days, the grapes will be so damaged that they can be used only for grape juice; in this case, the proceeds are only $1.30 per kilo.

The grape farmer can be sure of a harvest of 12,000 kilos. He chooses the risky second way of harvesting. Compute the advantages and the disadvantages that he faces compared with the first strategy.

Meteorologists computed that for every day in this two week period, the chance for rain is 15 percent. Compute the chance (probability) that it will rain in this two week period on more than two days.

The farmer chooses that way of harvesting which promises the largest expected yield. Which strategy will he choose? Illustrate your answer with a computation (HAVO, National Examination, The Netherlands, 1989. Used with permission.)

Properly constructed open questions, with a variety of short, long, and extended responses, offer possibilities for assessment at a level that is above the lowest level—whatever name we give to the lower levels. They may be called *knowledge outcomes, a variety of intellectual skills and abilities, computation and comprehension,* or *basic skills and facts.* Whatever the terminology used, it is generally agreed that we need other instruments like essay tests that provide a freedom of response, which is required for measuring complex or higher order learning outcomes.

*Essays*

An old-fashioned but seldom used tool in mathematics education is the essay test. As Gronlund (1968) stated: "Essay tests are inefficient for measuring knowledge outcomes, but they provide a freedom of response which is needed for measuring complex outcomes. These include the ability to create, to organize, to integrate, to express and similar behaviors that call for the production and synthesis of ideas."

The most notable characteristic of the essay test is the freedom of response it provides. The student is asked a question that requires producing one's own answer. The essay question places a premium on the ability to produce, integrate, and express ideas. The shortcomings of the essay task are equally well known: It offers only a limited sampling of achievement, the writing ability tends to influence the quality of the answer, and the essays are hard to score in an objective way.

Essays can come very close to being extended-response open questions, especially in mathematics. The snowplow problem could be considered an example of an essay, which brings us immediately to an often-mentioned aspect of the essay: Should it be done at school or at home? Usually the essay task is seen as a take-home task. However, this is not necessary; one can readily think of simpler essay problems that could be done at school but require a day (or so). An example given at some fifty schools in The Netherlands follows.

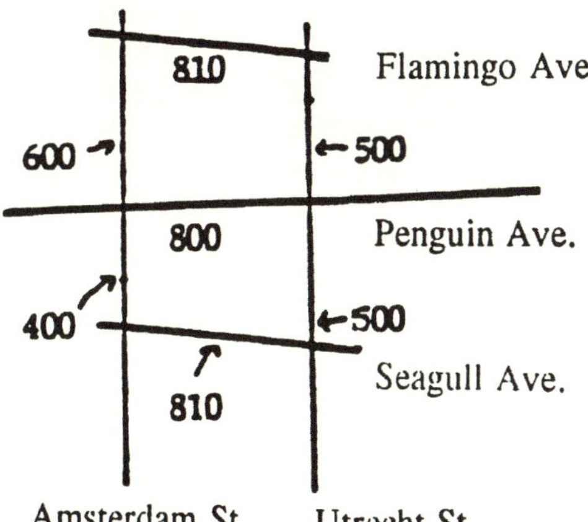

On the map, Amsterdam Street and Utrecht Street are the main streets with a maximum speed of 30 mph. At the intersections with Flamingo, Penguin, and Seagull Avenues are traffic lights. To get a smooth as possible traffic flow and to minimize irritation, it is considered important that the waiting time for red lights be minimized. If a driver has a green light, he or she should catch a green light when arriving at the next intersection, and so on for the next light. In this case we talk of a "green wave." To get as close as possible to a green wave situation, one has to consider for instance

—the duration of the green, yellow, and red light
—the relation between the different intersections
—the indication of advised travel speeds

It is known that a traffic light cannot be red for more than a 90-second interval and that each green period should last a minimum of 5 seconds. On average, the traffic on the main streets is four times as intensive as on the avenues. How would you adjust the traffic lights for each of the following more complex situations?

1. Bicyclists on Utrecht Street from north to south should have a green wave.
2. Both bicyclists and car drivers should have a green wave from north to south on Utrecht Street.
3. Both north and south traffic (cars and bicycles) on Utrecht Street should have a green wave as often as possible.
4. Both north and south traffic (cars and bicycles) on Utrecht and Amsterdam Streets should have a green wave as much as possible. Also take into consideration the traffic on Flamingo, Penguin, and Seagull Avenues: This traffic should have a reasonable flow too.

What basic principles and considerations should you use in general and in more complex situations? (National A-lympiade, The Netherlands, 1990. Used with permission.)

This example was meant for students at the upper secondary level, non-mathematics majors. The students worked in groups of four from 9 A.M. to 4 P.M. and were able to complete the task in a reasonable way. Note that the final question especially is an essaylike question; the other four are more or less extended-response open questions to make sure that the students get started in the first place.

Tasks like this can also be given individual students, at home or at school, depending on the goal that has to be measured.

It is generally accepted that the more precise or "closed" the questions are, the more objective is the scoring. Viewing this item from this perspective, one is tempted to conclude that this task can be scored with reasonable objectivity or better in a good intersubjective way. By this we mean that the scoring is done by two or more teachers independently. Although this may not be feasible practice for routine classroom teaching, it might be tried periodically with a fellow teacher. Essay tasks are typically not feasible for routine classroom teaching, so the incidental essay item might prove to be an excellent tool for creating interaction at the content level between teachers—an activity that is vital to good teaching at any school or institute.

It is interesting to compare this rather open problem with one that presents a similar context and problem but is meant for a timed-test situation at about the same age level and has to be completed within half an hour:

- Here you see a crossroads in Geldrop, The Netherlands, near the Great Church.

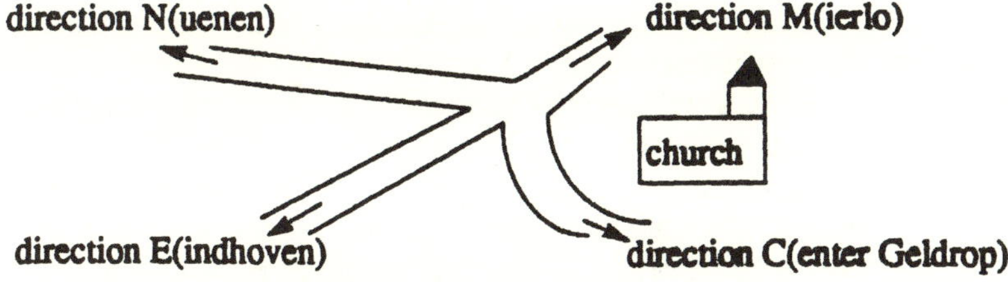


To let the traffic flow as smoothly as possible, the traffic lights have been regulated to avoid rush-hour traffic jams. A count showed the following number of vehicles had to pass through the crossroads during rush hour (per hour):

$$
A: \quad \text{from} \quad
\begin{array}{c}
M \\ N \\ E \\ C
\end{array}
\begin{pmatrix}
0 & 40 & 200 & 30 \\
30 & 0 & 80 & 50 \\
210 & 60 & 0 & 60 \\
30 & 40 & 80 & 0
\end{pmatrix}
\\
\quad\quad\quad\quad\quad M \quad N \quad E \quad C
$$

The matrices $G_1$, $G_2$, $G_3$, and $G_4$ show which directions have a green light and for how long. $^2/_3$ means that traffic can ride through a green light for a period of $^2/_3$ minute.

$$G_1: \quad \begin{array}{c|cccc} & M & N & E & C \\\hline M & 0 & \frac{2}{3} & \frac{2}{3} & 0 \\ N & 0 & 0 & 0 & 0 \\ E & \frac{2}{3} & 0 & 0 & \frac{2}{3} \\ C & 0 & 0 & 0 & 0 \end{array} \qquad G_3: \quad \begin{array}{c|cccc} & M & N & E & C \\\hline M & 0 & 0 & 0 & 0 \\ N & 0 & 0 & \frac{1}{2} & \frac{1}{2} \\ E & 0 & 0 & 0 & 0 \\ C & \frac{1}{2} & \frac{1}{2} & 0 & 0 \end{array}$$

$$G_2: \quad \begin{array}{c|cccc} & M & N & E & C \\\hline M & 0 & 0 & 0 & \frac{1}{3} \\ N & 0 & 0 & 0 & 0 \\ E & 0 & \frac{1}{3} & 0 & 0 \\ C & 0 & 0 & 0 & 0 \end{array} \qquad G_4: \quad \begin{array}{c|cccc} & M & N & E & C \\\hline M & 0 & 0 & 0 & 0 \\ N & \frac{1}{2} & 0 & 0 & 0 \\ E & 0 & 0 & 0 & 0 \\ C & 0 & 0 & \frac{1}{2} & 0 \end{array}$$

—How many cars come from the direction of Eindhoven (E) during that one hour? And how many travel toward the city center?

—How much time is needed for all lights to turn green exactly once?

—Determine $G = G_1 + G_2 + G_3 + G_4$ and thereafter $T = 30G$. What do the elements of $T$ signify?

—Ten cars per minute can pass through the green light. Show in a matrix the maximum number of cars that can pass in each direction in one hour.

—Compare this matrix to matrix $A$. Are the traffic lights regulated accurately? If not, can you make another matrix $G$ in which traffic can pass more smoothly? (van der Kooij, 1989).

It will be clear that we can easily turn this open question problem into an essay by simply asking the final question.

Essay tasks have long been neglected in mathematics education but certainly offer possibilities if carefully designed and if they can be evaluated in an appropriate way. Another classical tool that we use often as part of the teaching-learning process is the oral presentation or discussion.

*Oral Tasks*

In The Netherlands, oral examinations at the national level have a long tradition. Until 1974, an oral component was part of the national examination. Oral tasks still constitute one part of the examination and may be used by individual schools to constitute as much as 50 percent of the final score. There are many different forms, of which we cite three:

- an oral discussion on certain mathematical subjects that are known to the students in advance,
- an oral discussion on a subject—covering a take-home task—that is given to the students for study twenty minutes prior to that discussion,
- an oral discussion on a take-home task (or similar alternative task) after the task had been completed by the student (and scored by the teacher).

A global inventory of oral tasks at Dutch schools during the 1980s shows that the oral task was almost exclusively used to operationalize process goals. The solution, or product, per se was not important, and in many cases the examiners stopped the discussion short of the actual solution of the problem.

It has sometimes been argued that the students who perform well on restricted-time written tests do equally well on oral tasks. This would suggest that there is no point in going through the elaborate process of doing oral tasks. It appears that this effect is due partly to the fact that similar questions are asked on written and oral tasks. In this case, it can hardly come as a surprise that the results have a high degree of correlation.

We did some small studies that compared the results of different groups of students regarding the correlation between restricted-time written tests and oral tests—dealing, of course, with the same subject matter. One detailed study focused on two classes (twelfth grade) with twenty-eight students and two teachers who practiced intersubjective scoring. The correlation between the different results was 0.42, which seems to suggest that different learning outcomes were tested with the different formats (de Lange, 1987). At other schools, we carried out a similar study, although in less detail, and obtained a remarkably similar result: The correlation was 0.47, with seventy-three students involved.

It is not difficult to find protocols that "prove" that one can get notably different results in oral tasks, compared with written timed tests. Almost every teacher will have had that experience at some time in his or her classroom. It is also interesting to compare students' reactions to oral tasks: The results are sometimes surprising. One usually low-achieving girl did surprisingly well on oral tasks, surprising both the teacher and the student, who commented, "I am in favor of including oral tasks. When I get stuck with some detail in a restricted-time written test I usually get very nervous. This makes it impossible to solve the rest of the problem successfully. In this way I can try to prove that I really do understand the subject matter." Another student who was excellent in mathematics, according to his teacher and the results of written tests, was incapable of interpreting the results of a matrix multiplication during an oral test. This student said afterwards, "I am personally not really in favor of oral examinations. In my opinion, time is too restricted (total time per student was 20 minutes). There was not enough time to think things over because keeping your mouth shut may give the impression that you are trying to stretch time. Another drawback is the fact that you cannot do much computation. Finally, I don't like people watching and observing me. . . . "

The general conclusion of a larger group of students involved in oral tasks was that positive aspects outweighed negative aspects. Positive aspects of oral tasks that they mentioned explicitly were these: more questions on insight and theory and on mathematizing, less on arithmetic and computation; good atmosphere makes you feel more relaxed, because of hints you will never get "stuck"; not much attention to details, but more general questions. Negative points mentioned were pressure due to lack of time, pressure due to presence of officials, and little or no computation.

It will be clear from the preceding that oral tasks offer excellent opportunities for testing, but that they need careful preparation. The problems with oral tasks can easily be underestimated; one is tempted to take into consideration the perceived level of the student, ask questions similar to those on the written tasks, allow too little time, give help that is misleading (the student may mentally be on a track other than the one the teacher presumes). The teacher needs to take into account the fact that he or she has to spend much more time on oral tasks than the time spent with the student.

A completely different oral task than that just described, which we can categorize as "discussion" between the examiner and the student, is the "presentation" task. In this case, the student is asked to prepare a presentation about a subject to be discussed with the teacher. Those students planning to go on to higher education need some experience in presentation on mathematical subjects. Again, this area of learning and assessment is still, for the most part, undeveloped. The format can be employed very successfully if we include a discussion by the other students regarding the presentation, especially if we let the students "grade" the presentation: What are the strong and weak points and why are they strong or weak? Such discussion will in most cases lead to reflection by both the presenter and the teacher, as well as the audience, about the subject. This is a format that needs further exploration.

### Two-Stage Tasks

Any combination of different test formats may be termed two-stage tasks. An oral task on the same subject as an earlier written task is a typical example. Some years ago we explored two-stage tasks that basically consisted of one task carried out in two stages. The characteristics of this restricted two-stage task combine the advantages of the traditional restricted-time written tests and the possibilities that more open tasks offer. The characteristics of the first stage are that all students are administered the same test at the same time; all students must complete it within a fixed time limit; the test is oriented more toward finding out what students do not know than what they do know; usually the tests operationalize the lower goals, such as computation and comprehension; it consists of open questions; and scores are as objective as possible, given the fact that we exclude the multiple-choice format. These are then the characteristics of the first stage of the task.

The second stage should complement the first and address what we miss in the first stage, as well as the goals we really want to operationalize. The characteristics of the second stage are the following: The test has no time limit; the test is done at home; the test emphasizes what the student knows, rather than what he or she does not know; much attention is given to the operationalization of higher goals, such as interpretation, reflection, and communication; the structure of the test includes more extended-response open questions and essay type questions; and scoring can be difficult— intersubjectivity in grading should be stressed.

The test, as indicated earlier, consists of a problem with a number of questions. An example was the forester's test that we presented in part on page 108. The complete text is given to the students during a restricted-time session at school. They then have to read the entire test (around three pages) and decide which questions can be successfully completed in the classroom that day. In principle, they are free to tackle any of the questions but it came as no surprise that most students did in fact choose the open questions with short or long answers and left the essay questions for later. They were helped in making this decision by the fact that the first questions were also the more simple questions. When the bell rang, the students were asked to hand in their results, which, of course, were incomplete. When the scored tests were handed back to the students a week later, the scores were disclosed as well as the larger mistakes.

Now the second stage takes place. Given the teacher's feedback on the work they did, each student repeats and completes the work at home—without restrictions and completely free to respond to the questions as he or she chooses, whether one after the other, by way of an essay, or by a combination of these. After a certain interval, of perhaps three weeks, the students have to turn in the work again, and again scoring takes place. This sequence provides the teacher with two marks: a first stage (lower goals?) and a second stage (higher goals?) grade.

A total of thirteen questions on the test represented the formats as follows:

- open question or short answer: questions 1 and 2
- open question or long answer: questions 2, 3, 4, 5, 6, 7, 8, and 10
- essay or restricted response: questions 5, 8, 9, 10, 11, 12, and 13
- essay or extended response: questions 12 and 13.

Looking at this classification, one might expect students to handle successfully the first eight questions during the first stage. The remaining questions seem more appropriate for the second stage, at home. This was exactly what happened when this test was given its first trial in The Netherlands in two groups of twenty students each. The first seven questions were successfully answered during the first stage by more than 75 percent of the students; question 5 by between 50 and 75 percent, and the remaining questions were successfully handled by less than 50 percent of the students (with less than 25 percent on questions 9, 12, and 13).

It is not easy to convey a proper impression of the quality of the students' production in the second stage. The variety in their responses was enormous. Certain students completed some tasks in a straightforward way, answering the questions and paying no attention whatsoever to layout and related topics. Others turned in a veritable booklet, with color illustrations, self-made computer software, and typed or word-processed answers. Most students followed the order of the questions strictly and did not stray too far from their content. Some wrote an essay in which all of the questions were answered, and a number of students took the opportunity to show their own creativity in content-related questions.

Looking back at the results—as expressed both in scores and in student comments—we note the following:

1. There was a relatively wide spread in scores for the first stage—from very poor to excellent. In the second stage the students performed much better.
2. Girls' performance was relatively poorer than boys' in the first stage. At the second stage, this difference almost disappeared. In fact, the best results were from girls.
3. Intersubjective scoring was a satisfactory way to score the second stage.

The task of evaluating the second stage was given to thirteen teachers who did not know the students nor their results for the first stage; their marks for eleven students differed no more than 10 points on a 100-point scale.

### Production Tests

If one of our principles is that the testing should be positive, which means, first, that we should offer students the opportunity to use their abilities, and, second, that tests are part of the learning-teaching process, the students' own productions offer fine possibilities for achieving our purpose. The idea of own productions is not really new. Reports on experiences with this form of evaluation go back a long time. Treffers (1987) has introduced the distinction between construction and production, which is not a matter of principle. Rather, free production is regarded as the most promising way in which constructions may be expressed. By constructions, we mean (1) solving relatively open problems that elicit diverse forms of production, due to the great variety of solutions they admit, often at various levels of mathematization, and (2) solving incomplete problems that, before they can be solved, require the student to supply data and references.

The construction space for free productions might be even more extensive and include contriving one's own problems (easy, moderate, difficult) as a test paper or as a problem book about a theme or a course, authored to serve the next cohort of pupils (Streefland, 1990). For example, "Think of as many sums as you can with the result 3, or 5" (grade 1).

Grossman reported (1975) on the unexpected results achieved by this production task. She presents a few examples of work with first graders. We quote two and include the teachers' comments:

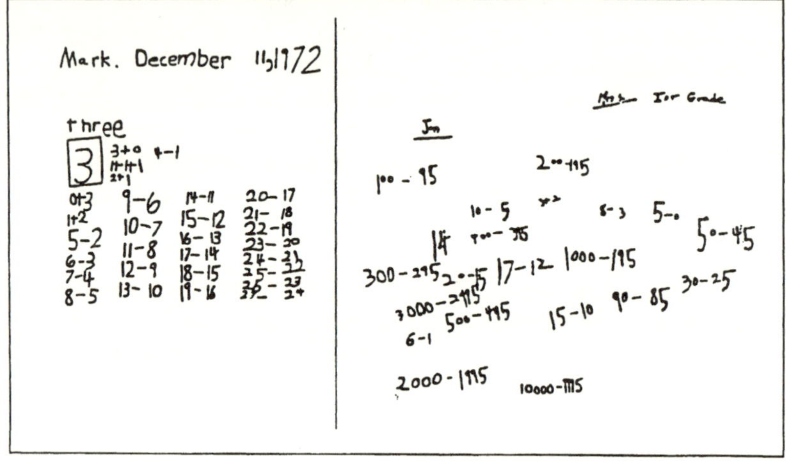

Figure 4.4   The response of two students to a free productions problem. (Streefland, 1990). Used with permission.

I knew Jon was bright because he understood so well all that I taught in my structured lessons, whether I followed the syllabus or went just a little beyond it. However, I never suspected that he could handle number combinations in hundreds and thousands. There I was, teaching combinations up to 20, limiting my expectations and the children's ceilings.

Mark was having trouble with arithmetic until I gave this assignment. He amazed me and he proved to himself that not only he could do arithmetic but that he couldn't stop doing it. (He handed in two extra papers on his own on subsequent days.) The other children loved the activity, too. My feeling was one of constant amazement that they could do it at all.

Streefland remarks that the teachers' comments show that both boys had amply exceeded the limits of the scholastic domain. Mark's work still reveals traces that show how he reflected on his activities. After a hesitant start, when he scanned the available arithmetic, he screwed up his courage, became conscious of self, wrote bigger, and sailed a fixed course through the system he built while constructing the problems. He transcended the boundaries of the arithmetic lesson and produced his own structure. At home, he continued intensively—the same Mark who was supposed to have problems with arithmetic. Jon must have seemed even more limited in his possibilities; yet, he anticipated a level of sums that were three grades higher in the curriculum—up to $10,000 - 9,995 = 5$. Like Mark, he worked systematically. Only in his written report was he a bit untidy.

Both of the boys reflected on what they had learned within the number system, and consequently they anticipated a future level of the teaching-learning process. A test item presented earlier— make as many sums as you can that equal 100 (page 104)—shows clearly that, on the one hand, we have an assessment tool and, on the other, we can use that tool for the learning process—a very important aspect for proper assessment.

We next address in some detail what took place in an American ninth grade on the subject of data visualization. The students had been working—suffering, interacting, thinking, discussing—for two weeks on a text designed in The Netherlands and based on the philosophy of realistic mathematics education. It is a philosophy, we believe, that fits reasonably well with the philosophy of the NCTM *Standards*. The Data Visualization booklet (de Lange & Verhage, 1992) was intended for about five weeks of class activities.

At the end of their second week, the students got their final test; the task was a simple one:

- At this moment, you have worked your way through the first two chapters of the book and completed a relatively ordinary test. This task is quite different: Design a test for your fellow students that covers the whole booklet. You can make your preparations from this point on. Look at magazines, books, and other available sources for data, charts, and graphs that you want to use. Write down ideas that come up during schooltime. After finishing the last lesson of this booklet, you will have another three weeks to design your test. Keep in mind:

1. The test should be taken in one hour.
2. You should know all the answers.

It is tempting to show many exciting examples of the students' production. To be honest, there were disappointments as well. One student simply took a reasonably well-chosen collection of exercises from the booklet, avoiding any risk or creativity. The next "higher" design strategy is the one that mathematics teachers often use: Take examples from the textbook, make small changes (in exponents, coefficients, or maybe context), and your test is ready. This worked for some students as well, although the answers sometimes made it painfully clear that the author was not among the brightest students.

Our first example shows an exercise that operationalizes the lower level:

- The following are the top 20 point leaders of the Edmonton Oiler team. Draw a stem-and-leaf diagram.

| | |
|---|---|
| WG 208 | DL 32 |
| JK 135 | KL 26 |
| PC 121 | PH 25 |
| MK 88 | KMC 23 |
| GA 81 | RG 23 |
| MN 63 | DJ 20 |
| MM 54 | DS 18 |
| CH 51 | LF 18 |
| DH 36 | BC 17 |
| WL 32 | JP 12 |

What is the average for the 20 pieces of data?
(4.13)

What is the median?
(2)

What is the difference between the average and the median?
(22,15)

Why is the average higher than the median?
(Some people had extremely high points.)

The next example is interesting because it tackles the problem of misrepresentation that often occurs in pictographs.

Two- or three-dimensional objects are used to represent one-dimensional facts. The subject, "fair" or "honest" graphs, got some attention in the booklet. One of the students created the following exercise:

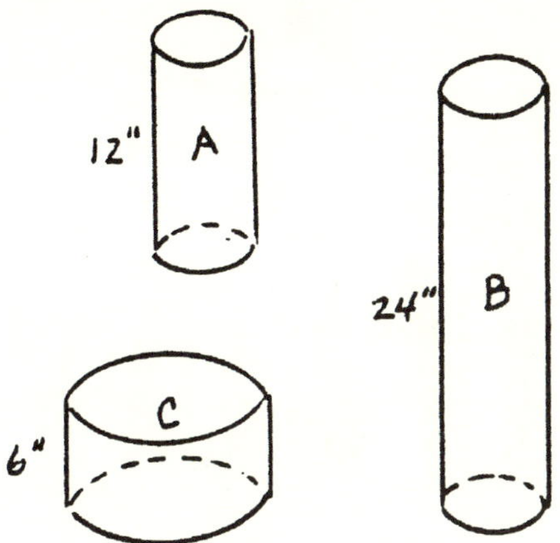


Does *B* or *C* show the volume of *A* doubled?

Which cylinder, *B* or *C*, has the same volume as *A*? Why? (de Lange, Burrill, Romberg, & van Reeuwijk, 1993. Used with permission.)

"Understanding Graphs" is a title that we took from one of the student tests. So far, we were mainly seeing rather straightforward questions about averages, means, and histograms, box plots and stem-and-leaf diagrams, and scatterplots. Some of the questions posed by the students already showed a tendency toward somewhat higher order thinking skills and some of them were surprisingly open. However, most of the questions discussed next are more about data visualization. As with the previous examples, we will not really discuss the problems but merely present them. It is well worth the effort to try to answer the questions and let the reader judge the quality of the problems designed by the students.

- The graph shown here is a bar graph involving information about how much money is being spent on military supplies compared to the country's GNP.

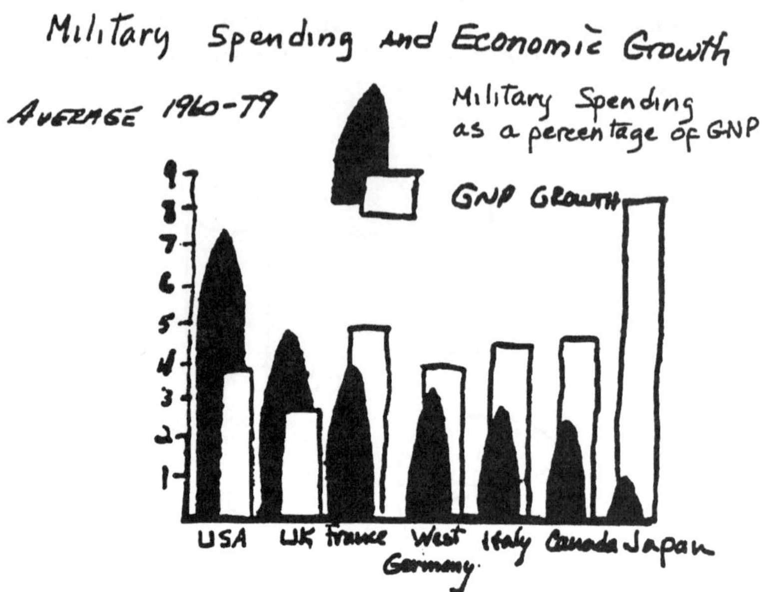



Is this information easy to read and understandable? Why or why not?
(No, this information is not easy to read because the numbers on the left have no meaning. One cannot tell if they mean millions, billions, etc.)
Could this information be represented better? Explain.
(No, because a line graph would not work, a box plot, pie graph, stem leaf, etc.)
Is this graph accurate? Explain your answer.
(No, because this graph is based on an average between 1960–1979.)(de Lange, Burrill, Romberg, & van Reeuwijk, 1993. Used with permission.)

The following example shows clearly that the student understood that the booklet was more than just an introduction to mean, median, box plots, and histograms.

■ In 1968, the United States was first and the Soviet Union was second in machine tool production. In 1988, Japan was first and West Germany was second. Compare the difference in dollars between the two sets.

(The difference in dollars in 1968 is roughly $3 billion and in 1988, it's $2 billion. This could mean that in 1988, the amount of the production of machine tools is more evenly spread.)

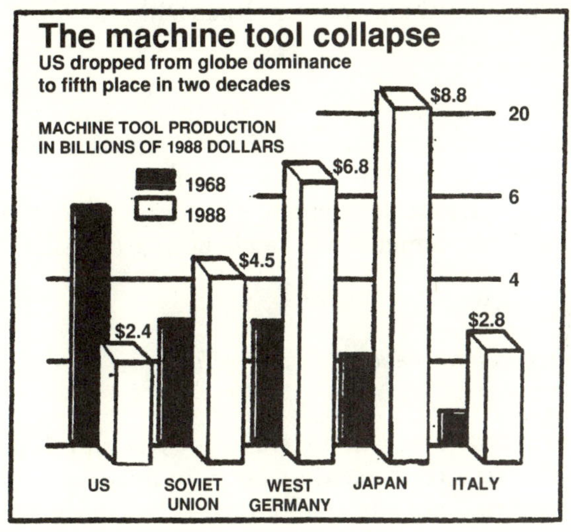


Why is data like this collected?
(It is collected to learn who is the leading producer of machine tools. For the United States, it is also helpful in that we know what to improve in our economy.)

Would a line graph represent this data better?
(No, I don't think so, because line graphs are primarily used to show sudden changes in connected things. A bar graph is better for this information because each country is separate and not connected in any specific way besides machine tool production.)

What is the average for each of the sets separately and combined?
(For both of the sets, the average is about $3.92 billion. It's only an approximation because you can't tell the exact numbers from the 1968 bars. For 1968 alone, it's about $2.74 billion. For 1988 alone, the numbers average $5.02 billion.)

> Is there any way to get the averages besides computing? (Yes, I think so because if you look at the length of the bars themselves and not the actual numbers, you can tell pretty closely for each of the individual sets. For both of the sets combined, it gets a little harder because you have to balance them on an $X$ sort of thing.) (de Lange, Burrill, Romberg, & van Reeuwijk, 1993. Used with permission.)

It will be clear from these examples that the students' own productions offer excellent possibilities for improvement for the teaching-learning process, as well as for assessment. It also goes without saying that there are difficulties and problems to overcome regarding the design, implementation, and scoring of tasks like those described.

### Fragmented Information Reasoning Test

A special kind of assessment was recently developed at the Freudenthal Institute to tackle a specific need in mathematics education. Mathematics in school has veered in the direction of increased applications, mathematization, and reasoning to use mathematical applications in other disciplines. College-bound students should be able to find relevant data from redundant information and should be able to cope with information or data that are lacking. They should be able to formulate a hypothesis and support this hypothesis in the best way possible, given the scant information. How does fact $A$ relate to fact $B$? This is a distinctly different activity than to give a mathematical proof. Proofs are relatively easy, certainly at secondary school level—if we teach the students any real proof at all, which means going beyond Pythagoras.

As we can see, too often the reasoning in many scientific publications leaves something to be desired. Especially in social and life sciences, "commonsense" logical reasoning is often lacking or incomplete. An effort to develop in students a capacity for this commonsense logical reasoning on a scientific level should be part of the mathematics curriculum. But how do we assess it? In response to this question we initiated the process that resulted in the first modest attempt to operationalize the goal of assessing reasoning.

Our main impetus for developing a means of assessing reasoning came from information presented in a scientific article entitled "The Importance of Aquatic Resources to Mesolithic Man at Island Sites in Denmark," by Noe-Nygaard (1982). This article had already been "translated" mathematically to a lower level because of the interesting use of secondary school level mathematics (de Lange, 1984).

The construction of the assessment instrument took the following course. We tried first to identify relevant information in the original article. Actually, we divided the article into several parts. We added other data; on the one hand, to give students some background, on the other, to make the information redundant. This resulted in twelve sets of information or data on twelve different sheets of paper. And the task for the four students involved was to support or reject the following hypothesis: "During the Mesolithic period, people from mid-Europe went farther north in spring and summer than previously assumed. Recent research leads to the conclusion that they did not stop in mid-Germany but went as far north as Denmark."

The twelve information sheets were very different in nature—one, for instance, simply provided data from an encyclopedia regarding the Mesolithic period; another, a map of Denmark; and the next taken almost directly from the original article: "Excavations near the now-vanished Lake Preastelyngen in Denmark resulted in finding numerous remains of fish skeletons. In total, around 5,000 parts were found. Scientific data methods have shown that these remains date from the end of the Mesolithic. Scientists have tried to identify which kind of fish were involved. According to these sources, surprisingly, many of the bones were from pike— around 80 percent of the bones. Pike was a popular fish at those times. The rest was prey fish like tench and perch."

It was expected that the students might get very confused when confronted with this test and might not know where to start. So a couple of activities were included at the lower level to make the format more useful. The original article offered ample opportunities for getting started, because not all of the graphic representations were without flaws. This resulted in the following page:

Because growing is a cyclic process . . . circular graphs are sometimes used to describe growth processes. The researchers at Praestelyngen Lake designed the following graphs that represent part of their findings:

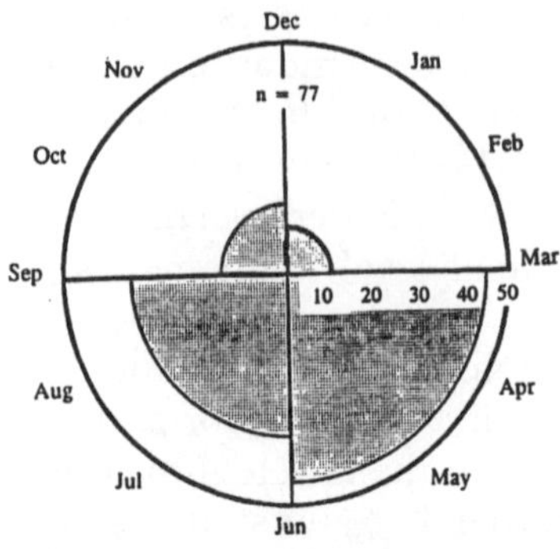
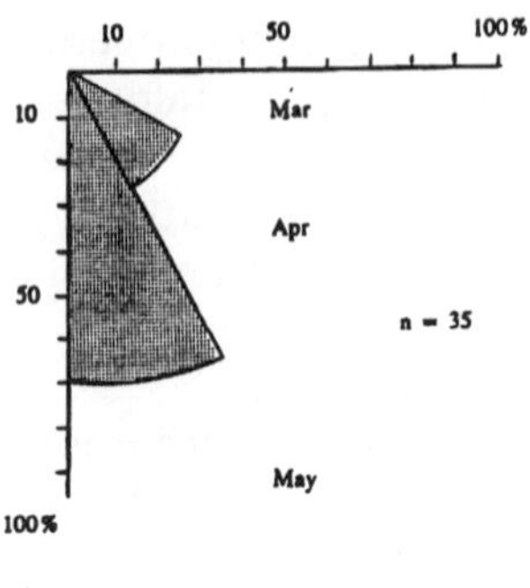


*Graph of the time of death of the pike at Lake Praestelyngen (in percentages).*

Some critical remarks can be made about the correctness of these graphs.

Study the graphs, keeping in mind the following questions:

How would the first graph have looked if the pike had all died in the period March–June? Do you get the same area for the dotted area?

And how about the second graph: Does the area for the month of May really represent the number of deaths in May?

Also give suggestions for better graphs and use those (if you like) in your work. (de Lange, for the National A-lympiade, 1992. Used with permission.)

The task was intended as a group task in a restricted time period. At the time of writing, only two experiments had taken place, each giving the students six hours.

The videotape of the first experiment indicates that all of the students operated in a similar way: First, they each read the materials and, after a short reflection on the nature of the task, they tackled the individual pages, starting with the page of graphs just shown. In this way, they slowly grew into the task. Next, they tried to glean the relevant information from each of the twelve pages. This was followed by a key activity: the discussion about whether or not to accept the hypothesis. It was very interesting to see a group of four male teacher-students work all morning resulting in a firm decision to reject the hypothesis. After a short lunch break, they decidedly chose that direction. One student took a more careful look at an atlas to get a clear idea of the geographical position of the places involved. After at least half an hour, the student at the atlas rose and told the other students with much confidence: "They have been there!" An interesting discussion now took place as to whether they should change and support the hypothesis. They did change. And here we have an essential aspect of this kind of test: There should be room for two directions. The students can either support or reject the hypothesis.

We are still far from reaching firm conclusions about this test format. But indications are that it could be promising: The students reason in a way that we seldom see: They have to look at details, discuss which position to take, write and communicate at an appropriate level, design and use adequate illustrations, and at the same time, carry out a substantial number of activities at lower levels. One question to be explored is whether we can design similar activities at different levels.

*Project Work*

Project work has had supporters for a long time, especially in the United Kingdom. However, it is still unclear what a project really is. Several definitions have been used, including the following: "A project should be a well-balanced and well-presented piece of work, showing completeness and depth of study, and offering genuine mathematical involvement with the topic" (Allen, 1977). As a result of the ambiguities in some of the terms used in this definition, a more formal definition of a project cannot have much value.

An essential question is Why should we have projects at all? This is even more relevant because many tests do not meet our standard that a test be practical: An appropriate project is not only difficult to design (like many of the formats just described), but it is also hard to organize and carry out. Allen (1977) gives several arguments in favor of project work: In the process of completing a project students should be extending their experience in either the collection of information or learning techniques to handle information. He adds that naturally we cannot specify the precise knowledge to be gained in the form of objectives because of the potential variety of subject matter open to student selection.

Another goal of project work can be the original application of a technique or a number of techniques that require the student to reflect upon the structure of the problem. He or she will also have to consider alternative models and finally make some kind of evaluation. So it seems there is little question but that the ideal situation should operationalize lower, middle, and higher level goals, irrespective of the taxonomy we use. Another important issue is whether student involvement in the project will produce results different from those expected in other settings. On the basis of our research, we know that we do indeed measure goals in project work that are different from goals in traditional course work. Allen reports a correlation around 0.44, which seems to indicate that project work results are more or less independent of the more traditional assessment work.

Project work can be done in groups, and usually is, but it also can be done individually. It can be carried out according to a strict scenario or in complete freedom. It can be rather short, or it can be a lengthy investigation. Subjects for investigation are abundant: the adjustment of traffic lights (an example used earlier), determining which supermarket is cheapest, figuring out how to keep records so that a store never has too much or too little in stock, tracking pupil and student movement in school, placement of new buildings in a town layout, determining the pollution of lakes or air, finding the blind spots in traffic control, making maps and projections, and in many data-collecting activities.

It is difficult to describe a typical project activity. A structured example may require an entire week of work on a special book or subject that otherwise never would have been treated. When students have an entire week in which to integrate mathematics with other information, their freedom is somewhat limited, but the result for both the students and the teacher is much more easily judged than in projects that leave them greater freedom.

At this time, it is unclear how we can best assess project work. There is little doubt that well-designed project work (which means, in most cases, small projects) will have beneficial effects for the teaching and learning process. However, we cannot be certain whether project work measures this process more effectively, or whether it measures different dimensions of it, compared with some of other formats we have described. Project work should be considered as a viable assessment tool only if the other formats described, which are usually easier to execute, do not operationalize the goals we wish to test.

### Portfolios

In the United States, much attention has recently been devoted to portfolio assessment in mathematics. Portfolios may seem comparable to projects, but are actually quite different. Murphy and Smith

(1990), in describing the use of portfolios in the teaching and learning of writing, state the following: "Portfolios are obviously more than a collection of artifacts. They are, even before the unfilled folders are home from the store, a reason for talking. And, depending on the way the talk goes, the portfolio can take many different forms. In a sense, coming up with a portfolio project is like choosing what to teach. The decision automatically creates possibilities and limitations. In the infinite scheme of what can be taught, teachers choose for their particular classroom communities. In the same way, they can make decisions about portfolios with themselves and their students in mind."

Mumme (1990) argues that a portfolio—as a tool for assessment—focuses on the student's productive work. It measures what the students can do rather than what they cannot do—one, of course, of our primary principles. But there are many other instruments that do this.

Portfolios provide insight into many aspects of student growth—in mathematical thinking, understanding, ability to express ideas, attitudes toward problem solving, and others. A portfolio, when used for assessment, is more than a "folder," according to Mumme. It is a deliberate collection of student work that can be used to provide evidence of understanding and accomplishment in mathematics. A portfolio offers the potential of providing more authentic information than other formats about a student's mathematical endeavors. Such information can help students assess their mathematical progress, assist teachers in making instructional decisions, improve communication with parents, and enable educators to assess the mathematics program at their school.

To support her plea for portfolio assessment, Mumme quotes *Everybody Counts* (MSEB, 1989): "We must ensure that tests measure what is of value, not just what is easy to test. If we want students to investigate, explore, and discover, assessment must not measure just mimicry mathematics. By confusing means and ends, by making testing more important than learning, present practice holds today's students hostage to yesterday's mistakes."

Does a portfolio really make manifest the philosophy implicit in the MSEB's position? To answer this question, it is necessary to consider the general attributes of portfolio assessment, described earlier, in light of the following examples of student mathematics activities that can be included in a portfolio (listed in Mumme's recent monograph, *Portfolio Assessment in Mathematics* [1990]):

> student written work
> individual and group work
> rough drafts and finished products
> student writing
> projects and investigation
> diagrams, graphs, charts
> work written in the student's first language
> photos of student's work
> audiotapes of student explanations or oral presentation
> videotapes
> computer printouts and disks
> work dealing with the same mathematical ideas sampled at different times

According to Mumme, teachers like portfolios for discussion between teacher and student, for meetings between teachers, for discussion between students, and for presentation to the school and community. The question remains, Are portfolios the best tool for this purpose? As one who has seen alternative forms of assessment functioning in different places on a larger scale, I would like to pose some questions about the effectiveness and validity of portfolios—interpreted in the broadest way. The impression I have from, for example, frequent visits to the United States is that portfolio use is gaining momentum because current assessment procedures are not effective measures of teaching and learning. Portfolios offer freedom, are fun, are not structured, are open; and they represent a lot of work for teachers—exactly the opposite of established assessment practices. But many teachers are unaware of the variety of assessment formats available. And many of those formats do the same things that portfolios are supposed to do for students, teachers, and administrators in a much more reliable and direct way.

### Balanced Assessment Package

Overall, according to Burkhardt and Resnick (1991), for teachers to judge student performance and growth, a balanced assessment package is needed. Their claim is that such a package represents the range of mathematics that we now aim for students to be able to do, as articulated in the NCTM *Standards* (1989). The package these authors propose, based on a sequence of mathematics activities lasting a little over four weeks, is well worth careful study in light of this review of assessment instruments. Those who examine this package will decide for themselves how well balanced it is. But it seems certainly more balanced than is usual in assessment. And it poses the possibility that even more balanced assessment packages can be created. The question thus becomes, Who decides what constitutes appropriate balance? We have described numerous formats that attempt to operationalize objectives for specific levels and outcomes. But deciding on the makeup of an ideal package depends not only on research outcomes that are currently lacking, but also on a philosophy of mathematics and mathematics education that is in the process of evolving.

Burkhardt and Resnick claim that their package aligns with the *Standards* (NCTM, 1989). Mumme (1990) claims that portfolios offer opportunities to operationalize the philosophy of the *Standards*. The states of California and Connecticut, as well as several others, claim that they offer innovative forms of assessment that fit the *Standards*. We believe that much of the work we have done in The Netherlands on assessment tasks is consistent with the principles of the *Standards*. It may be that many others working in the field of assessment will make the same claim, and we have to be concerned that the efforts of some will converge on the practical tasks, which, we already know from our research, seldom measure middle and higher learning goals.

## SOME FINAL NOTES ON CHANGE AND ASSESSMENT

### Points That Require Attention

Designing tests and administering them in ordinary classroom practice is not a simple task, as we have seen. During the design of assessment tasks, one has to be very clear which goals are being operationalized, which context to choose or formats to consider, and the practicality of presenting it in the classroom. But other points also need serious consideration when choosing a balanced package of assessment tools, as the following questions indicate:

- Is the test to be taken within a fixed time interval (restricted-time test)?
- Is the test to be taken individually or in groups?
- Is the test to be taken at home or at school?
- Is the test a single-strand test, an integrated test, or an interdisciplinary test?

- Is the test part of a continuous assessment practice, or is it part of a more discrete scenario?
- And for some people the most important question, How objectively scorable is the test and what scoring tools do we have to assure that the scoring will be as accurate and fair as possible?

These questions are deserving of a separate article. The following reflections are based on our experience during a decade of experiments on assessment.

*The Timed Test Can Have Many Variants.* We usually identify the timed test with the individual restricted-time written test at school. This format forces students to perform at the same sitting under the same external conditions within the same time frame. Research (de Lange, 1987), as well as informal teacher experiences, seems to indicate that girls perform less well than boys under these conditions.

But timed tests are also tests where the students are allowed to work in groups or at home if we pose strict time limits on their task. In these relatively open situations, it is more difficult to decide how much time we should allow students. The greater freedom provided by these options offers more time for reflection and creativity.

Restricted-time written tests have certain advantages, like practicality, and disadvantages, such as the need on the part of the student for peak performance at a time decided by the teacher and under pressure. But the reduction of time restrictions in tests has disadvantages too, especially if the consequence is that the students can take their tests home. One option for carrying out assessment tasks in an almost unrestricted-time format is to do it at school: Have the students do an extensive performance task from 9 A.M. to 2 P.M. or so. Of course, practically, this poses real problems, given the standard school day in most schools, but if we take into consideration the fact that this arrangement is necessary only once or twice per school year, it may improve the picture considerably.

*Certain Tasks Are More Suited for Group Work Than Others.* It may be enlightening to give a multiple-choice task to a couple of students and ask them to reason aloud. Of course, multiple-choice tasks are seldom suited for group work, but as a learning experience they can be very rewarding. Not infrequently, the discussion takes a course like the following: "This is certainly wrong, this too. So there are two possibilities. What does the designer of the tasks want us to answer?" As a Dutch teacher once commented, "The students are secondhand thinkers. They are not accomplishing a mathematical activity but merely reflecting on past experiences with multiple-choice and trying to follow the thinking process of the designer." Another teacher expressed explicitly what many people know but some test designers deny: "Teachers really teach how to pass a multiple-choice test. A large part of the year is spent in training students for the test instead of teaching them mathematics."

Group work is not very compatible with multiple-choice, but offers good possibilities in other formats. A group can be just two students, sometimes three or four, or even the whole class. Two students can work very successfully on extended-response questions, essays, two-stage tasks, production tasks, and in a limited way, on project work. For projects, it is usually necessary for several persons to cooperate successfully, but individual projects are feasible too. The fragmented information reasoning task has to be done by several students working together because the discussion about the hypothesis is the kernel of the whole activity. There is no way for the students to avoid this stage.

The advantages of group work in assessment are well known: reflection on one's own thinking, reasoning and reflection, communication, production, cooperation, arguing, negotiating. The disadvantages are also clear: evaluating individual contributions fairly (if such are made), the practicality problem, the makeup of the groups, and the scoring. Cohen (1986) argues that evaluation is not so difficult as it may seem: If the students are at least fourth graders, many important questions can be answered via a questionnaire. But the questionnaire that she refers to focuses on how the students evaluate the group process, not on the mathematical quality of the result.

Another factor to be considered in group work is the prospect it offers for systematic interactive scoring by an outside observer. Cohen argues that it is relatively easy to obtain a rough estimate of the rates of participation of the students. This may not be true at all: A student who is thinking on his own may, to an outside observer, seem to contribute little, yet still come up with the turning point in the discussion, and this point may not always be easily recognized. Besides this complication, we have to bear in mind that much group work in assessment has to take place at home or outside of school.

We have observed successful group work in assessment in different ways: strongly organized groups with a presider, secretary, and workhorses of different kinds. The students were evaluated on the basis of two factors: how they functioned in the group (the assessment Cohen is talking about) and their individual mathematical contributions separately. Another option is to allow the group process to work by itself. The group as a whole is responsible for the product and all of the individual members of the group get the same score, irrespective of their contributions. The underlying assumption is that the harder working or smarter students will not let the others coast along for a free ride and that everyone has to make a contribution. This seems to work especially well in groups of two students. It seems important to note at this point that because of the assumption that we are working with a balanced package, we have no problem with part of the final score for a student being gathered in individual tasks and part in group work.

Group work can take place within the classroom or school, but also at home. This holds for other tasks as well. The two-stage task was especially designed with the idea of combining restricted-time school-based tasks with a more unrestricted part to be done at home. Although the task was difficult to design, the results were very promising, as we indicated earlier in this chapter. Because the task remained unchanged, the components to be included were clear—only the conditions were different. On the one hand, we had limited time, a school context, a pressure factor, and equal conditions for the students (at least in the externals); and on the other, unlimited time, minimum pressure, and access to additional sources both in materials and in persons. The results for both students and teachers were satisfying because of the two grades and the insight it seemed to offer on the different qualities of the students on the same mathematical content at different levels. The fact that the correlation between the results of the two phases was low (< .50) makes it necessary to consider the "take home part" of a balanced package seriously.

*Test Breadth.* Another delicate subject that needs more attention than it usually gets is the problem of the "broadness" of a test. All too often, the students have to prepare for an algebra test, or a geometry test. Sometimes they may prepare for an algebra or geometry test, but in most cases they deal with an assortment of algebra or of geometry items and are not challenged to integrate them. It seems important to invest some effort in developing integrated tasks that draw on all of the domains students are supposed to be familiar with or know how to handle at a particular moment. This is not a simple challenge, especially if we continue to label the different aspects of mathematics as we usually do and offer students compartmentalized mathematics.

Another area that needs further study, experimentation, and careful evaluation is test preparation. "To prepare for a test" is a well-known and often well-defined activity for students. And in the standardized test arena, there is even a multimillion dollar industry that claims credit for doing this successfully. This, of course, is unfortunate for education and for politics. Assessment is a part of the teaching-learning process; it will be interesting to see how well the principles and goals for mathematics assessment, as published in *For Good Measure* (MSEB, 1991), will be implemented in the year 2000. The first principle articulated coincides with our own first principle: "The primary purpose of assessment is to improve learning and teaching." And to make things a little clearer: "Whether with classroom assessment or external assessment, the process and result of assessment must inform and enhance the learning and teaching process rather than narrow or restrict it." If this ideal is to be realized, we also have to make decisions about the continuity of the process of assessment. Do we really want our students to "prepare for a test," or should a student, in theory, be ready for a test at any time? To select the assessment tools for a balanced package includes the spread in time of the use of these tools. Do we do all assessment during a single week in a semester or spread it out evenly over time, and what are the different advantages of one approach or the other? In our opinion, assessment should be a more or less continuous activity—like the teaching-learning process—for improving that process in an optimal way.

Finally, there is the matter of objective scoring. Our extensive classroom work, research (de Lange, 1987), and commonsense thinking make the following hypothesis look very easy to defend: The gains we make by obtaining a more or less complete measure of overall knowledge and capabilities by using a balanced package of assessment will by far outweigh the disadvantage that we have by "losing" a complete objective score. Intersubjective scoring and proper scoring instructions give enough guarantees for a fair measure, fair to the student and fair to the curriculum. Admittedly, we need more information and further trial studies and research. But the development of new assessment tools and guidelines on how to use them and score them is essential. Successful reform in mathematics education requires that curriculum, its philosophy, methods of instruction, teacher training and enhancement, and assessment be revitalized in tandem. The purpose of this chapter is to anticipate the problems and aspects of the assessment that we must address in the near future.

*Assessment: No Change Without Problems*

A number of national documents, the changes that are taking place in mathematics education, and our decade of research on the teaching and learning of school mathematics all point to the necessity of primary changes in the way student knowledge of mathematics is assessed. Assessment should utilize real problems, which often will mean real-world problems and their applications, as well as the complexities that result when the world is brought into the classroom. Having made this point, it is immediately necessary to look at a number of other issues that we have to confront. These include the following:

- Teachers, test designers, parents, administrators, public officials, and citizens need to develop a new attitude toward assessment. This point is often underestimated. We cannot simply rely on assessment summits, reform publications, and other compelling developments in assessment. Society has a certain image of assessment that will take years to change and improve. The damage done by past assessment practices, especially in the United States, will take at least a decade to repair, and no quick or cheap solutions are available.
- Different levels of mathematical activities need different assessment tools, which are demanding and time consuming to design and require a major research and testing effort.
- To design a balanced assessment package will be difficult.
- To interpret the different strategies and processes that the students will come up with in more open assessment will be hard for teachers. Teacher training with special emphasis on assessment is not only needed, but will enrich teachers' understanding of the problems we are dealing with.
- Different problems need different contexts, which take into account all kinds of variables, a point made earlier. A special problem will be to find the balance between a good context and a good mathematical problem.
- Scoring and judging the quality of the diverse assessment formats available to us will be more complex and varied and will be considered more difficult than scoring and judging test items is at present.

If these issues come as a surprise to a certain degree, then it only makes clearer the magnitude of the challenges we face; it defines the seriousness of the situation in assessment at the present time. Assessment has become separated from its major participants: the students, their teachers, and the mathematics curriculum. There can indeed be no change without problems.

## REFERENCES


Allen, R. A. B. (1977). *Project work in mathematical studies.* Glamorgan: United World College of the Atlantic.

Bodin, A. (1993). "What does to assess mean?" In M. Niss (ed.), *Investigations into assessment in mathematics education: An ICMI Study.* Dordrecht, The Netherlands: Kluwer Academic Publishers.

Boertien, H. (1990). *Afsluiting basisvorming. Voorstellen voor afsluitende toetsen wiskunde* [Examples of intermediate assessments in basic secondary education]. Arnhem, The Netherlands: National Center for Educational Evaluation, CITO.

Burkhardt, H., & Resnick, L. B. (1991). *A balanced assessment package.* Nottingham, UK: Shell Centre.

Chernak, R., Kustiner, L. E., & Phillips, L. (1990). "The Snowplow Problem." *UMAP Journal, 11,* no. 3:241–276.

Cockcroft, W. H. (1982). *Mathematics counts: Report of the commission of inquiry into the teaching of mathematics in schools.* London: Her Majesty's Stationery Office.

Cohen, E. G. (1986). *Designing group work.* New York: Teachers College Press.

de Lange, J. (1979). "Contextuele problemen." *Euclides* 55.

———. (1984). "Viskunde voor allen." *Nieuwe Wiskrant 4,* no. 2.

———. (1987). *Mathematics, insight and meaning.* Utrecht, The Netherlands: OW & OC.

———. (1991). *Flying, meaningful math.* Scotts Valley, CA: Wings for Learning.

———, Burrill, G., Romberg, T. A., & van Reeuwijk, M. (1993). *Learning and testing mathematics in context—the case: Data visualization.* Scotts Valley, CA: Sunburst/Wings for Learning.

———, & Verhage, H. B. (1992). *Data Visualization.* Scotts Valley, CA: Sunburst/Wings for Learning.

Freudenthal, H. (1983). *Didactical phenomenology of mathematical structures.* Dordrecht, The Netherlands: Reidel.

———. (1991). *Revisiting mathematics education.* Dordrecht, The Netherlands: Kluwer Academic Publishers.

Galbraith, P. L. (1993). "Paradigms, problems and assessment: Some ideological implications." In M. Niss (ed.), *Investigations into assessment in mathematics education, an ICMI Study.* Dordrecht, The Netherlands: Kluwer Academic Publishers.

Gravemeijer, K., van den Heuvel-Panhuizen, M., & Streefland, L. (1990). *Contexts, free productions, tests and geometry in realistic mathematics education.* Utrecht, The Netherlands: OW & OC.

Gravemeijer, K., van den Heuvel-Panhuizen, M, van Donselaar, G., Ruesink, N., Streefland, L., Vermeulen, W., te Woerd, E., & van der Ploeg, D. (1992). *Methoden in het reken wiskundeonderwijs, een rijke context voor vergelijkend onderzoek [Textbooks in mathematics education, a rich context for comparative research].* Utrecht, The Netherlands: Freudenthal Institute.

Gronlund, N. E. (1968). *Constructing achievement tests.* Englewood Cliffs, NJ: Prentice-Hall.

Grossman, R. (1975). "Open-ended lessons bring unexpected surprises." *Mathematics Teaching 71.*

HAVO, Mathematics A. (1989). *National Experimental Examination.* Utrecht, The Netherlands: Freudenthal Institute.

———. (1990). *National Experimental Examination.* Utrecht, The Netherlands: Freudenthal Institute.

———. (1991). *National Experimental Examination.* Utrecht, The Netherlands: Freudenthal Institute.

Johnston, W. B., & Packer, A. E. (1987). *Workforce 2000, work and workers for the twenty-first century.* Indianapolis: Hudson Institute.

Kindt, M. (1979). *De reis om de wereld in 80 dagen [Around the world in eighty days].* Utrecht, The Netherlands: IOWO.

Lapointe, A. E., Mead, N. Z. A., & Phillips, G. W. (1989). *A world of difference: An international assessment of science and mathematics.* Princeton, NJ: Educational Testing Service.

Maassen, J., & Verhoef, N. C. (1990). "De eerste experimentele wiskunde A-lympiade." *Nieuwe Wiskrant 9*, no. 4.

Mathematical Sciences Education Board (MSEB). (1989). *Everybody counts: A report to the nation on the future of mathematics education.* Washington, DC: National Academy Press.

———. (1990). *Reshaping school mathematics: A philosophy and framework of curriculum.* Washington, DC: National Academy Press.

———. (1991). *For good measure: Principles and goals for mathematics assessment.* Washington, DC: National Academy Press.

MAVO, Mathematics A. (1991). *National examination.* Utrecht, The Netherlands: Freudenthal Institute.

McKnight, C. C., Crosswhite, F. J., Dossey, J. A., Kifer, E., Swafford, J. O., Travers, K. J., & Cooney, T. J. (1987). *The underachieving curriculum: Assessing United States school mathematics from an international perspective.* Champaign, IL: Stipes Publishing.

Mumme, J. (1990). *Portfolio assessment in mathematics*. Santa Barbara: University of California.

Murphy, S., & Smith, M. A. (1990). "Talking about portfolios." *National Writing Project—Center for the Study of Writing Quarterly* [Berkeley, CA].

National Council of Teachers of Mathematics (NCTM). (1989). *Curriculum and evaluation standards for school mathematics*. Reston, VA: Author.

National Mathematics A-lympiade. (1990). Under the auspices of Nederlandse Onderwijs Commissie voor Wiskunde van het Wiskundig Genootshap [Dutch Mathematics Education Committee of the Mathematical Society]. Utrecht: Freudenthal Institute.

———. (1992). Under the auspices of Nederlandse Onderwijs Commissie voor Wiskunde van het Wiskundig Genootshap [Dutch Mathematics Education Committee of the Mathematical Society]. Utrecht: Freudenthal Institute.

Niss, M. (1992). "Assessment of mathematical applications and modelling in mathematics." Paper presented at ICTMA 5 conference.

Noe-Nygaard, E. (1982). *The importance of aquatic resources to mesolithic man at island sites in Denmark* (Internal paper). Copenhagen: University of Copenhagen.

Oaxaca, J., & Reynolds, W. A. (1988). *Changing America: The new face of science and engineering*. (Interim report). Washington, DC: Task Force on Women, Minorities, and the Handicapped in Science and Technology.

Oldham, E. E., Russell, H. H., Weinzweig, A. I., & Garden, R. A. (1989). "The international grid and item pool." In K. J. Travers & I. Westbury (eds.), *The IEA Study in Mathematics 1: Analysis of Mathematics Curricula*, pp. 22–23. Oxford: Pergamon Press.

Pea, R. D. (1987). "Cognitive technologies for mathematics education." In A. H. Schoenfeld (ed.), *Cognitive science and mathematics education*, pp. 89–122. Hillsdale, NJ: Lawrence Erlbaum Associates.

Phillips, D. C. (1987). *Philosophy, science, and social inquiry*. New York: Pergamon Press.

Popper, K. (1968). *Conjectures and refutations*. New York: Harper Books.

Querelle, N. (1984). "Ach Ja." *Nieuwe Wiskrant 4*, no. 1.

Resnick, L. B. (1987). *Education and learning to think*. Washington, DC: National Academy Press.

Rheinboldt, W. C. (1985). *Future directions in computational mathematics, algorithms, and scientific software*. Philadelphia: Society for Industrial and Applied Mathematics.

Romberg, T. A. (1991). "How one comes to know: Models and theories of the learning of mathematics." In M. Niss (ed.), *Investigations into assessment in mathematics education, an ICMI Study*. Dordrecht, The Netherlands: Kluwer Academic Publishers.

Stevenson, H. W., Lee, S. Y., & Stigler, J. W. (1986). "Mathematics achievement of Chinese, Japanese and American children." *Science* 231:693–699.

Stigler, J. W., & Perry, M. (1988). "Cross-cultural studies of mathematics teaching and learning: Recent findings and new directions." In D. A. Grouws, T. J. Cooney, & D. Jones (eds.), *Perspective on research on effective mathematics teaching*, pp. 194–223. Reston, VA: National Council of Teachers of Mathematics.

Streefland, L. (1990). "Free productions in teaching and learning mathematics." In K. Gravemeijer, M. van den Heuvel-Panhuizen, & L. Streefland (eds.), *Contexts, free productions, tests and geometry in realistic mathematics education*. Utrecht, The Netherlands: OW & OC.

———. (1991). *Fractions in realistic mathematics education.* Dordrecht, The Netherlands: Kluwer Academic Publishers.

Travers, K. J., & Westbury, I. (1989). *The IEA Study of Mathematics I: Analysis of mathematics curricula.* Oxford: Pergamon Press.

Treffers, A. (1987). *Three dimensions. A model of goal and theory description in mathematics education.* Dordrecht, The Netherlands: Reidel.

———, & Goffree, F. (1985). "Rational analysis of realistic mathematics education." In L. Streefland (ed.), *Proceedings of PME-9.* Utrecht, The Netherlands: OW & OC.

van den Brink, J. (1987). "Children as arithmetic book authors." *For the Learning of Mathematics, 7,* no. 2.

van den Heuvel-Panhuizen, M. (1990). Realistic arithmetic/mathematics instruction and tests. In K. Gravemeijer, M. van den Heuvel-Panhuizen, & L. Streefland (eds.), *Contexts, free productions, tests and geometry in realistic mathematics education.* Utrecht, The Netherlands: OW & OC.

———, & Gravemeijer, K. P. E. (1990). *Reken wiskunde Toetsen Groep 3.* Utrecht, The Netherlands: OW & OC/ISOR, University of Utrecht.

van der Kooij, H. (1989). "Het eerste Hawex-examen." *Nieuwe Wiskrant 9,* no. 1.

VWO, Mathematics A. (1987). *College preparation, Secondary education. National Examination.* Arnhem, The Netherlands: National Center of Educational Evaluation, CITO.

W12–16, team (1991). Achtergronden van het nieuwe leerplan Wiskunde 12–16, band 1 en 2. Utrecht/Enschede, The Netherlands: Freudenthal Institute/SLO.

# 5 ❖ The Invalidity of Standardized Testing for Measuring Mathematics Achievement

*Robert E. Stake*

Mathematics teachers across the United States report strong and still growing emphasis on standardized achievement testing in their schools.[1] They recognize a need to distinguish between valid and invalid uses of testing. The purpose of this chapter is to help them make that distinction.

Testing as an activity and individual tests as a tool are neither valid nor invalid until the results are interpreted in some way. Only the interpretation of test scores in particular situations can be said to be valid or invalid (American Psychological Association, 1985; Cronbach, 1980; Jaeger & Tittle, 1980; Linn, 1989; Messick, 1989).

Standardized mathematics tests are used in many situations to obtain valid indication of which students are better at solving the types of problems included in the test. Student motivation for scoring well (Raven, 1992), time limits of the test (Shohamy, 1984), unfamiliarity of the language and format of the test, and other features of test production and administration (Traxler, 1951; Airasian & Madaus, 1983; Freeman et al., 1983) can contribute to invalidity in interpretation. But good achievement tests, properly administered, with scores cautiously interpreted, are an appropriate component of mathematics education. In our schools, it is a responsibility of teachers to recognize and reward superior performance. Test scores contribute to teacher awareness of how students compare with one another.

Standardized mathematics test scores are not, however, a sound basis for indicating how well students are becoming educated in mathematics. Scores that do a good job of indicating which students are doing best and which are doing relatively poorly do not necessarily provide a valid indication of subject-matter mastery. One test alone will not provide valid measurement of the mathematics achievement of individual students or of a group as a whole. Test content almost always is too narrow. Just as a few students do not represent all the students in a school and a few books do not represent all the books in a library, twenty or thirty test items do not represent the broad range of mathematics skills and knowledge that teachers are teaching. For measurement of subject matter attained, the simplicity of testing is at odds with the complexity of teaching and learning.

When we talk about mathematics education, we refer to long-standing and well-deliberated definitions, though they are not often put into words. With good reason, we are reluctant to change our notions of education to fit the simpler definitions offered by standardized testing. For example, part of the established meaning among educators and others is that education is a personal process and a personally unique accomplishment. For each student, experience in and out of school is different; thus, the formal and informal meanings of arithmetic, algebra, geometry, and all of mathematics are different from student to student. Furthermore, each mathematics teacher's understanding is personally constructed and somewhat tuned to the cultural experience of teachers. The ideas they share in the classroom and stimulate in the minds of students differ from teacher to teacher. Of course we speak of standard courses of study, common objectives, and shared understanding of mathematics—but the education of youth includes the application of mathematics to many *unique* experiences of past and future. Part of the invalidity of achievement testing is due to constraints imposed by common aims. But there are other constraints as well.

Much of mathematics education is beyond accurate assessment. That does not, of course, relieve the teacher of responsibility for searching for valid evidence that education is happening. Much good evidence of mathematics achievement comes from reflective interaction with individual students and within the class as a whole—interaction during recitation, exercises, projects, and testing. Test scores alone are a flimsy indicator of the mathematics that students have learned.

In this chapter, I will describe the mismatch between the highly abstract but simple constructs of mathematics held by developers of standardized tests and the situational and much more compounded conceptualization of mathematics held by teachers—a difference that extends, of course, to interpretations of mathematics achievement. This is a problem common to other subject matters but only mathematics education will be considered here.

The teacher-reader will be reminded that standardized testing serves certain school administration and classroom management purposes. The appropriateness of such use is not considered directly to be an issue of validity, yet it belongs in the discussion. Because validity is rooted in the interpretations teachers and others make of test scores, I think it important to show how testing as a process, how assessment as an instrument of educational reform, is seen by teachers to contribute to or detract from school improvement. The more they acquiesce in the idea that good test scores mean good teaching, the more readily they will regard testing as related to good changes in school programming and the more likely they will interpret tests as defining the mathematics to be achieved.

For any particular use of a standardized achievement test, validity of the measurement indicates the quality of information conveyed about how well the students are achieving—for lifetime so far, for the year, or even for the lesson. The key distinction in this chapter is between a generic, single dimension concept of achievement—a view promoted by test specialists—and a complex, experiential conceptualization of achievement detailing the many steps, the many differentiations, of content and skill—a view held by teachers. My conclusion will be that these views are so different that the panorama of achievement that mathematics teachers regularly scan cannot be measured validly with the standardized achievement tests in use today.

## MATHEMATICS EDUCATION

It seems we all know what school mathematics is. For many people, such common experience seems to need no definition. But that presumption, held too by more than a few mathematics teachers, is a misperception. Mathematics educator Thomas Romberg has noted, "Mathematics is viewed as a vast collection of vaguely related concepts and skills to be mastered in strict order, with the sole objective of becoming competent at carrying out some algorithmic procedure in order to produce correct answers on sets of stereotyped exercises" (1987). Mathematics indeed embraces a vast array of concepts and operations, but in the teaching of mathematics, student understanding is more the objective than development of powers of calculation. In point of fact, the detail and the scope of mathematics education exceed the best of definitions.

### The Underperception of Mathematics Teaching and Learning

Neither good nor bad teachers stick to the point. In the classroom, good teachers roam the content terrain, point out and extend major connections, introduce concrete situations of relevance. As they teach them, mathematics knowledge and skill are not collections of discrete elements. Each problem type and algorithm is linked into various networks of knowledge, diverse traits, other systems of thinking (Scheffler, 1975; Romberg and Carpenter, 1986). One-digit addition and two-digit addition are closely linked; one-digit addition and two-digit multiplication are less close, yet linked in several ways. These several ways become multitudinous when applications are acknowledged. The applications of mathematics quickly become too numerous to itemize in tables of contents, lists of objectives, lesson plans—yet the teacher, not only deliberately but subtly and unconsciously, adds dimensions of meaning to each operation and concept.

The chapter titles of a mathematics textbook seem simple enough. For example, the chapters of the book used by the upper sixth grade in the Duxbury (Massachusetts) Intermediate School in 1989 are listed in table 5.1. A quick review of chapter 1 ("Addition and Subtraction of Whole Numbers") finds further subdivision into the topics of place value, reading and writing groups of three numerals, one- and two-digit addition and subtraction, properties of addition, three- and four-digit addition, money units, missing numbers, five-digit addition, three-digit to six-digit subtraction; subtraction with zeros, comparing numbers, greatest and least numbers, rounding numbers, estimation of sums and differences, Roman numerals—with some special attention to consumer skills, career interests, and problem solving. And each of these topics could be further subdivided. The inventory of topics spanning all fifteen chapters is extensive.

TABLE 5.1. CHAPTER TITLES OF A MIDDLE SCHOOL MATHEMATICS TEXTBOOK

| | |
|---|---|
| 1. Addition and Subtraction of Whole Numbers | 9. Probability |
| 2. Multiplication and Division of Whole Numbers | 10. Statistics and Graphing |
| | 11. Ratio, Proportion, & Percents |
| 3. Decimals | 12. Measurement |
| 4. Multiplication and Division of Decimals | 13. Perimeter, Area, and Volume |
| 5. Geometry | 14. Integers |
| 6. Factors and Multiples | 15. Using Triangles |
| 7. Addition and Substraction of Fractions | |
| 8. Multiplication and Division of Fractions | |

(From Vannatta & Stoeckinger, 1980. Used with permission.)

Teachers classify mathematics just as the textbook authors did, into various domains, sometimes for teaching and testing (Collis, 1982; Collis & Watson, 1989).[2] Their formal classification headings are useful as conceptual structure but draw too much attention to the well-known main topics of mathematics education. Educational researcher Lauren Resnick has observed, "The range of mathematical concepts to be learned [is] much broader and only a few of these crafts have been intensively studied." (1989, p. 164). Furthermore, learnings within a class or even within a small subclass are related to each other in many more ways than indicated by any single classification scheme. And many of the tasks and concepts within a subclass have important uniquenesses. To show this, we can examine the seven items of figure 5.1, seven problems not unlikely to appear in a 40-minute activity in algebra class.

---

1.    $20 \times 1.8 + 32 = ?$        2.    $(1/5)(20)(9) + 32 = ?$

3.    $\dfrac{2 \times 20 \times 9 + 32}{10} = ?$        4.    $y = 1.8x + 32$. Solve for $y$ if $x = 20$

5.    Convert 20°C to Fahrenheit. $F = (9/5)C + 32$

6.    Convert 20°C to Fahrenheit. $C = (5/9)(F - 32)$

7.    Ann wants to know today's temperature on the Fahrenheit scale. Her thermometer reads 20 degrees Celsius. What is the Fahrenheit temperature?

---

Figure 5.1.  A family of seven mathematics items.

These seven items cut across several content domains (Hively et al., 1973), yet a teacher might include all of them within a single lesson, within a single objective, or refer the solution of each of them to a single page of the textbook. Each item is unique, a special variation on the others. Each will be more or less well understood by students and thus more or less difficult. Still, different in form and notwithstanding transformations, they belong to a family. The family here is not defined by mathematical operations as much as by the practical problem of dealing with two temperature scales, Celsius and Fahrenheit.

*Inventories*

In the backs of our heads, we all have epistemological inventories of mathematics education that extend beyond the families, classes, levels, and lattices of quantification. These inventories are particularly broad when we include the applications of mathematics. The inventories can be organized around any one of the many conceptual structures proposed, such as that by Edward Haertel and David Wiley (1990; see also Henderson, 1963). Such classification schemes gravitate toward a powerful simple structure. Few of them reflect the real complexity of mathematics education to be found at every grade level. Inventories of teaching and learning, were they actually recorded, would show that complexity. They would reveal each teacher's complex conception of the nature of mathematics education.

For our inquiry into test validity, we need content inventories, not just categories of topics and principal kinds of mathematical activity, but inventories that reflect existing definitions of education and the complexity of teaching and learning in existing classrooms. Even people who know little mathematics can identify numerous categories. But teachers go much further, particularly in their choices of what and how to teach. Their conceptualization of the content of mathematics education is vast and detailed.[3] They decide, for example, whether to treat vertical addition the same way they treat horizontal addition. Most treat subtraction with borrowing as different from subtraction without borrowing. Some would treat multiplication of decimal numbers with zeros immediately following the decimal point as a special learning. Mathematics educators have been diligent in classifying such operations, but teaching practice regularly creates a host of additional subclassifications.

1.    Three children wish to divide two oranges evenly among themselves. Carefully peeled, one orange is found to have 12 sections, the other 13. What should they do?

2.    Three children wish to divide two oranges evenly among themselves. Unpeeled, one orange weighs 8 1/8 ounces and has 12 sections. The other weighs 7 3/4 ounces and has 13 sections. What should they do?

Figure 5.2. Two problem-solving exercises.

Consider the problem-solving exercises in figure 5.2. What if two sections of an orange are withered? And what if juice is lost by cutting? At what point does inequality not matter? We pause reflectively before placing these two exercises in the same category in our inventory. As teachers, we associate teaching strategy with learning tasks: Should we schedule peer group dialogues (Easley & Easley, 1992) and cooperative learning (Johnson & Johnson, 1991)? Is it the right time to refer to spherical geometry? In examining the invalidity of testing in this chapter, I will raise questions of logic, pedagogy, learning activity, difficulty, and utility. Already it should be apparent that it is not reasonable to suppose that any one task *represents* its category adequately. Can one suppose that a student's performance on one item will reveal how that same student would perform on another?

The many transformations of a mathematics problem extend beyond mere restatement, examples of which are shown in figure 5.1, into multidimensional extension. The transformation of tasks is further exemplified by the five questions in the previous paragraph. The various forms and language of a teacher presentation are part of what is learned by the child. Children have a considerable capacity for recognizing item type and transformation. Capacity grows as experience grows. The teacher contributes not only by drawing attention to performance tasks, but by engaging students in expository discourse about both large and small tranformations.[4]

In the mathematics classroom, such transformations arise, over and over, minute by minute. Some are simple, some are complex (Scheffler, 1973; Romberg, 1987). Though many transformations are deliberate, mathematics teaching takes the envelope of transformation largely for granted. So do students, parents, administrators, and policy makers. All find *simple* ways of representing inventories of operations and tasks. The labels people use for identifying the domains or topics or families of mathematics items suggest a homogeneity and generality that lead us to summarize performance by a single test score. Test scores seriously understate the diversity and complexity of teaching and learning. Mathematics education, then, appears to be more coherent and simply structured than it is.

*The Artificiality of Mathematics Achievement as a Construct*[5]

It is not artificial for a closely observing teacher to describe how well a student has worked a mathematics exercise or project. It is not artificial to indicate how many problems or test items were answered correctly. It is not artificial to conclude that a student has achieved a level of mastery, at least for the time being, over a particular group of exercises. But, to generalize broadly about achievement is artificial. It is common for people to treat mathematics achievement as real[6]—but risky. It is artificial and risky to conclude that a student has achieved proficiency over a type of exercise such as addition of fractions or the binomial theorem. It is artificial and risky to allude to achievement of a content so vague as sixth grade mathematics.

In speaking of mathematics achievement, one alludes to a selection of mathematics to be taught or that has been learned. As indicated in the previous section, I choose to call such a selection an *inventory*.[7] An infinite number of mathematical tasks exists. For any one moment, just what mathematics are we talking about? The selection is identified by the inventory. The inventory will not include "all of mathematics," much of which even the brightest mathematician does not know. We are thinking of all of the mathematics of interest just now. It could be all the mathematics taught in this school in sixth grade or all the mathematics needed for a student to be reasonably qualified when entering an engineering program at the state university. Inventories need not be a consciously itemized but must have substance, structure, boundaries, and lots of detail.

If the concept of mathematics achievement is to be useful, the inventory of the mathematics potentially achievable needs to be to some extent delimited and realized. Does the inventory include multiplication of fractions, the simple uses of a hand calculator, applications of the Pythagorean theorem, a notion of the work of Bertrand Russell, orienteering? Specification of the contents of the inventory need not be formal; it can be experiential and intuited—though, if there is to be meaningful dialogue about mathematics achievement, there should be some shared meaning of the inventory.

Each inventory may relate to goal statements, textbook exercises, and item pools. Subsidiary domains and boundaries are inevitably inexact. Each formulation of mathematics learning, each formalized inventory, is an umbrella for a vast array of tasks, habits, skills, and knowledge. Such formulations are large understatements of the mathematics intuitively and properly included in the practicing inventories of teachers.

If there were strong interdependencies among the many domains of mathematics, the need for an elaborate inventory would diminish. Were advanced skills simply composed of "prerequisite" skills, as claimed by Gagné (1967), the inventory could be easily specified. Were knowledge of calculus derivable from knowledge of trigonometry, we could use the latter to indicate the former. How much of a definition of education can be derived? When we have a well-developed set of relationships about the construction of an entity and can directly measure some of the parts, we can calculate other characteristics. The derived measures are not artificial.

But, in spite of its reputation as a logically coherent aggregate of knowledge, in spite of the common view that advanced mathematics skills are determined by prerequisite skills, we have no set of formal relationships that bind together the many domains of mathematics and mathematics achievement. Not only does understanding simultaneous equations remain largely independent of understanding permutations, but even the knowledge of fractions and the knowledge of decimals remain largely independent—a fact that leads me to presume that no universal system is possible. For now, at least, the best epistemological relationships we have are few, partial, and hypothetical. It is obvious, for example, that successful long division requires some competence in subtraction, but mastery of many subtraction problems cannot be assumed, given mastery of the main types of long division. The field of mathematics is too complex and dissociated to permit interpretation of one aspect of mathematics achievement on the basis of knowledge of another aspect.

### The Dissociation of Mathematics Topics

A person has many mathematics knowledges and skills—each being simultaneously used and forgotten, forever incomplete. Many knowledges and skills are related but few are highly interdependent, their dependence complicated by their incompleteness. When mathematics is considered in the broadest sense, a surface of personal achievement stands near zero in many places and rises irregularly and not very predictably in others.[8]

As mentioned before, we seldom are thinking of all mathematics. We usually limit our thoughts of mathematics achievement to those things covered by certain goals or particular chapters or teaching of a specific grade in a specific school, a terrain much more circumscribed. In spite of its reputation as highly integrated—that is, as a succession of prerequisite learnings—the topics of mathematics are quite dissociated. The same is true of mathematics education. Generalizing from one aspect of achievement to another is problematic. For example, though many people comprehend both, understanding symmetry remains independent of understanding orthogonality. Often our concern is with the content of mathematics and not with some generalized notion of mathematics ability. The content associated with a goal or contained within a lesson is often too heterogeneous for a few items to represent the rest.

Often our representations, our talk, does not require precision. There are times when all we want is a rough indication of how much a youngster has achieved. A teacher's recitation questions, chapter tests, and midterm grades, for example, serve as rough indicators of achievement. These approximations are bolstered by teacher knowledge of what has happened in the classroom. Achievement is assessed with reference to an inventory. (For outsiders, and even for the students, much of the teacher's inventory of the mathematics covered remains vague.) Among experienced teachers especially, the mathematics content is shared through custom and conversation. Still, precision of assessment is out of reach.[9] Each teacher's inventory is different. *It is important to recognize that as used by even the most knowledgeable teachers and testing specialists, the concept of mathematics achievement is artificial and imprecise, suitable at times for casual reference but a questionable basis for indicating how much mathematics a student knows.*

The concept of mathematics ability, sometimes deduced from performance on achievement tests, suffers from the same lack of common inventory of the mathematics covered. Ignoring content, the construct, mathematics ability, is useful as an indication—relative to other learners—of how much learning time or teaching effort will be required in subsequent courses. Relative standings remain quite stable for a fixed group of students, stable even as individuals pass on into other comparable groups.

Notwithstanding useful predictions, the way the concept of mathematics ability has been contrived has been injurious. Many test specialists, especially those advocating an item-response-theory approach to assessment (Hambleton, 1989), force disparate aspects of mathematics into a single indicator, omitting from their definitions those tasks and forms of knowledge that do not fit their scales nicely. The injury occurs when teachers, failing to see certain topics included in the tests, drop those topics from the inventories to be taught. Topical items most useful for predicting mathematics achievement are not necessarily good for defining it. Mathematics ability and mathematics achievement are not interchangeable concepts.[10]

When it is necessary for us to estimate, to generalize over unknown terrain, to presume the nature of the whole, to work with entities the contents of which have not been specified, our measures of the whole are artificial. Except in the simplest situations, the formal measurement of mathematics achievement is artificial.

### Teacher Conceptualization of Mathematics Education

As every teacher knows, there are shortcomings in education in America—including teacher shortcomings. If ours were an excellent educational system, thousands of teachers presently teaching would not be teaching.[11] But the fact of the matter is that the teachers we have are one of the stronger assets of the system, much stronger, in my opinion, than the administrators, the textbooks, the willingness of students to be students,[12] and the tests.[13] All could be better, of course. And at times, each of us believes, "If I were to yield to the pressures to change, conditions would become even worse." We are more or less locked into a mediocre educational system and appalled at the prospect of further deterioration. The tests that show our national achievement to be poor are not wrong but, as part of the problem, drive schooling toward standardized, authenticated mediocrity.

Current teachers are an asset mainly because they have a long-developed and far-reaching conceptualization of the connections[14] of ideas and behaviors that constitute a certain high school course or the year long lessons for a particular elementary school grade (Lieberman, 1984; Connell, 1985; Lampert, 1988). An experienced mathematics teacher has a strong idea of what topics should be covered, the calendar and time allotments involved, the relationship and interdependence of topics, the nuances and subclassifications of topics, diverse applications of topics, the relevance of topics to standardized testing, opportunities for enrichment and cooperative learning, nurturing independent thinking and self-directed learning,[15] ways of increasing motivation and decreasing discouragement, what the stumbling blocks will be, how socialization and conflict preempt academics, what experience students bring to the classroom, the expectations of students, parents, and other teachers. The work of teaching is complex.

I want to emphasize again the complexity of mathematics education. Somewhere in the mind of each mathematics teacher an inventory of the topics to be taught exists. Each topic, alone, is as intricate as a tree, with large and small branches, with twigs and lacy buds and leaves, individually dispensable but collectively vitalizing the tree. The parts are comprehensive and capable of personal interpretation. Trunk and main limbs are represented by the classification of goals, objectives, chapters, and types of problems—extremely important as structure and discipline for the emergent learnings of mathematics. But the connection of mathematics to experience (playing store or calculating water needs for a camp-out) to other subject matter and preparation for livelihood depends on the teacher's acquaintance with limbs and foliage. More than anything else, I believe, it is the teacher's comprehension of the subject matter that distinguishes between professional work and "functionarial" extension of the district curriculum office.

The district curriculum office has a critical role to play and many mathematics coordinators play it well. Individual teacher conceptualizations need to be nourished and protected; and in the case of those conceptualizations of teachers that are misguided or error ridden, redirected. The district coordinator supports the strong teachers and looks for ways to make students less dependent on the weak ones. The competent coordinator works to assist teachers in resisting the impairment of their conceptualizations by standardized testing.

These conceptualizations of mathematics on the part of teachers and particularly the inventory of mathematics to be taught are the critical epistemology of education. Comprehensiveness, integrity of content, and topical uniqueness are no longer certain to be found in the minds of superintendents, in the presentations of textbooks, or in the coverage of tests. No one is reading John Dewey. Few mathematics teachers know a mathematician with whom to discuss their field. Authority in subject matter is being preempted by the syllabi, the textbooks, and the tests—each with a leaning toward simplification, increasingly myopically bent on raising (potentially embarrassing) achievement scores. The situation is in flux. Within the context of that still vital autonomy exercised in most classrooms, the complexity of teaching continues. It draws upon that repository of mathematics education—content and technique—in the minds of the teachers, an impoverished cache, but a precious asset.

*Teacher-Made Assessments*

In this last decade of the twentieth century, education remains labor intensive. Efforts to automate teaching have been largely unsuccessful.[16] Why do we continue to put at least one expensive laborer in each classroom of every school? It is not because teacher unions are featherbedding. It is not because it takes a scholar to maintain discipline. It is not primarily to assure the choice and presentation of subject matter. We place almost 3 million teachers in American classrooms because managing the conditions for learning requires constant attention—recognizing readiness to learn, the uniquenesses of students, obstacles and intrusions; perceiving "progress within wrong answers," a never-ending assessment of student achievement.[17]

Following the ordinary practices of schooling, assessment summaries by teachers are not used to inform school administrators or the public of the scholastic integrity of the school nor to provide parents an understanding of career prospects for their children. Such *reporting* of educational progress is beyond the current skill of teachers, at present and in the future. As described repeatedly in the research literature (Gage, 1972; Giroux, 1988; Lortie, 1975; Resnick, 1989; Rosenshine, 1970), teachers use their ineffable, informal assessments to direct the activities of learners, reallocating time on task, recognizing patterns of idiosyncratic thinking, modifying interpretations. It is a form of assessment based little on a science of education, attendant little to formal testing; it is rather an intuitive artistry that matures in the reflective experience[18] of day-to-day teaching. Artistry rather than technology prevails because education, not mere training, is a highly individual experience. Many governors, newspaper writers, and educators claim that more of teaching should be decided centrally, in advance, and standardized across classrooms.[19] Even in the decentralized school, to a certain extent, major school goals are prespecified and common. In all schools, teaching varies from room to room for good reason: Each school is different; each teacher is different; the children are different. We educators and researchers back away from the state of the art when we support blanket prescriptions for heterogeneous schools and youngsters. Our formal plans are embarrassingly simplistic when compared to the routine and intuitive conceptualizations held by teachers. Teacher-made assessments are essential. To organize each course of study to fit national specifications would be more than a revolution; it would overthrow all of the serious notions of education we currently hold. Teaching has developed as an art;[20] as a technology, it is far less sophisticated.[21]

*Representing Education*

As all people do, teachers use simple representations. Their course outlines and lesson plans briefly list topics and activities. To satisfy the requirements of administrators or to talk to visiting parents, they sometimes refer to lists of objectives such as those articulated for the state of Georgia, abbreviated in table 5.2. But in thinking of how and what they will be teaching, teachers work at a much higher level of complexity.

TABLE 5.2. GOALS FOR EDUCATION IN GEORGIA (Excerpted)

The Georgia Board of Education has adopted student goal statements that identify the ideal skills and attitudes a graduate of Georgia's educational system should strive to achieve through instructional programs in the state public schools. The State board believes that the instructional program in the public schools should provide each individual with opportunities to develop abilities so that he or she

communicates effectively
uses essential mathematics skills
recognizes the need for lifelong learning
has the background to begin career pursuits
participates as a citizen in our democratic system

[And for mathematics in Georgia:] The mathematics section of the Quality Core Curriculum consists of objectives relating to concepts, process skills and problem solving at each grade level, kindergarten through eighth. In grades 9–12, objectives are given for each mathematics course. Mathematics began and continues to be a way of organizing one's world, through the study of quantity and space, their properties and the relationship[s] within and between these concepts. Mathematics is first experienced as a language created to describe the world, accompanied by rules that govern its use.

[And for Algebra I, the Topics/Concepts are identified as:] E. Polynominals
13. Identifies polynomial expressions
14. Adds and subtracts polynomials
15. Uses of laws of exponents necessary to perform polynomial operations

Complexity of teacher thinking was illustrated earlier in figure 5.1 by seven mathematics items. To a testing specialist, these items are points on a single scale; they measure essentially the same thing. To a teacher, each is unique. The statistical correlation among the seven would run high, but each item requires its own understanding of terms and operations. The mathematics teacher extends instruction to the details of each item.

Getting any six of the items right does not assure getting the seventh right. To a teacher, mathematics achievement is not a matter of getting the best score on the test, but of understanding and performing the work.

Within mathematics education, far more interweaving and interdependence of meaning occur than is apparent on a list or content-behavior grid (J. Wilson, 1971, p. 646). What if we tried to represent the similarity (proximity) and sequentiality (directions) of mathematics topics? The hypothetical map in figure 5.3 might stimulate our thinking. Here, for example, the topics of trigonometry appear closer to geometry than to arithmetic. If we had more detail, we would expect to see *percentage* lying closer to *fractions* than to *probability*. Such a two-dimensional map raises many questions but turns out to be as unsatisfying as a list. The relationships overwhelm the mapping.

Yet, when we analyze what a teacher is doing, we find topics and activities connected in logical ways as if all were mapped there in the teacher's mind. When we ask for a representation, the teacher seldom produces a detailed guideline as to what teaching fits where. Indirectly more than directly, the teacher has transformed complex epistemological relationships into a course schedule and on-the-spot action and reaction. When we analyze the thrust, we find teaching not aimed at developing some general mathematical ability but at

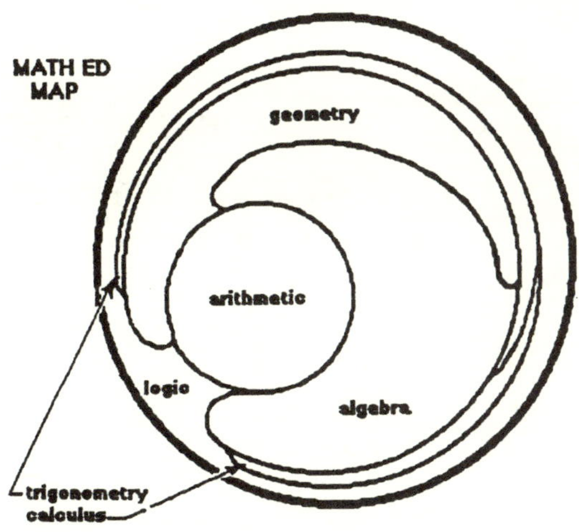


Figure 5.3. A map of mathematics education content.

developing a knowledge of specific topics and skills in solving specific kinds of problems. The inventory is the tacit map by which the pursuit of knowledge is rationalized.

Mathematics teachers incorporate anticipated student behavior into instruction. They allocate a large portion of time to operations and problem solving. Their conceptualization of mathematics teaching is process oriented more than it is outcomes

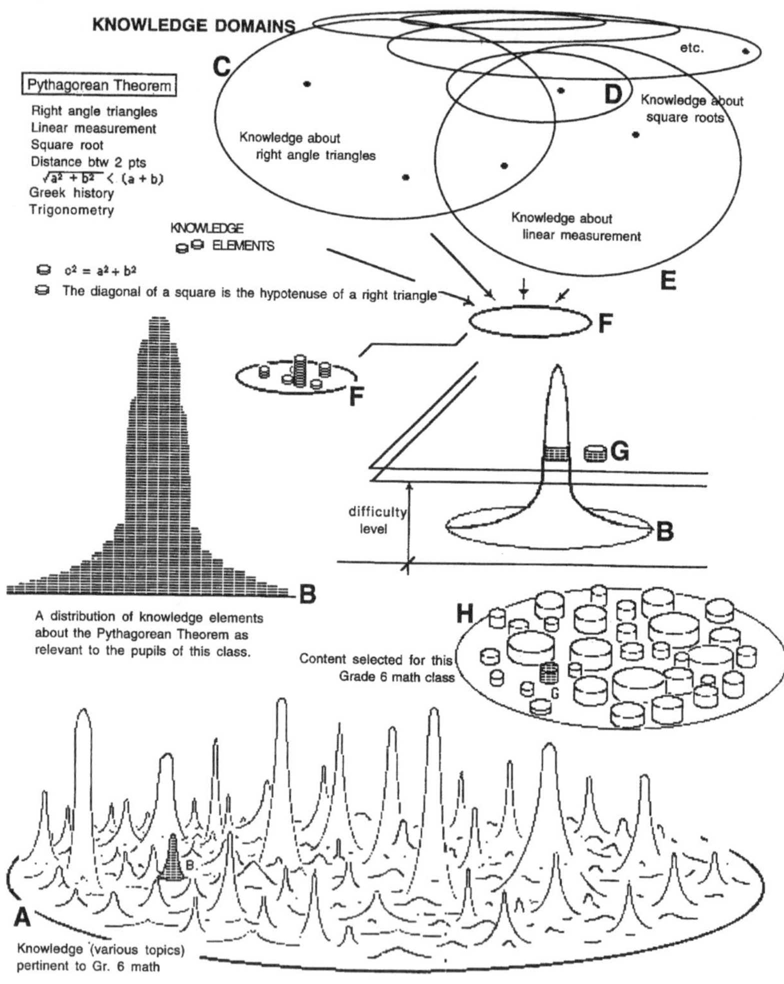


Figure 5.4. An impressionistic representation of a teacher's approach to teaching the Pythagorean theorem.

oriented. The teachers strive for high quality experience, immersion in the topic, honing the particular operation. Few mathematics teachers think first of making children "numerate" or (unless harrassed) pumping for better scores on an achievement test. Their first aim is to help children gain command of a far-reaching, unspoken inventory of subject matter, outlined perhaps as the National Council of Teachers of Mathematics (1989) proposed but extending to a network of detail as salient itself as the major classifications.

How does a sixth grade teacher approach the lesson? Figure 5.4 is my impressionistic representation of choices made by a teacher as to what to teach on a certain day about the Pythagorean theorem.[22] The topic is mentioned in the state list of learner objectives and identified in the district curriculum guide and the textbook the teacher is using. To a degree, the textbook author defines what will be taught but, especially in recitation, the teacher modifies course content to fit the situation, noting especially the frame of mind of the present student group. Reflecting on the many pertinent topics for the class, (topography A, bottom of the figure), the teacher considers the facts, concepts, relationships, and applications of the Pythagorean theorem. Distribution B represents a closer look at what is most relevant for these particular sixth graders. The teacher draws elements from several knowledge bases (circles C, D, and E) to obtain a small selection to teach (plate F), then thinks about learning difficulty (cylinder G) and inserts it as content (tray H) for this class.[23] The teacher anticipates a small presentation with graphics, reading, seatwork, and homework. When he or she presents the Pythagorean theorem in class, ideas are modified as the conversations of instruction occur, shaped, of course, by the teacher's overall conceptualization of mathematics education.

The naiveté of figures 5.3 and 5.4 is obvious; no less so that of the lists of tables 5.1 and 5.2. Graphic technology to *represent* pedagogy and epistemology is not highly developed.[24] Classification systems and content-skill grids are common in curriculum offices but there are few devices to represent the conceptual links between topics and guide pedagogical moves from one content to another. Yet, just as ancient travelers reached destinations before there were maps, teachers teach without maps, build without blueprints. Intuitively, good teachers merge topical paths, capitalize on personal experience, and draw out and preserve the students' lines of thought.

And let us project these graphic representations further to include the teacher's recollection of student achievement. We note first that most students are remembered primarily as members of a class that experienced the scheduled instruction and engaged (to an intensity varying across and within students) in mathematical tasks. During the school term the class was exposed to the teacher's inventory of mathematics pertinent to that term. Both class and inventory are remembered as ordinary in certain ways and unique in certain ways. The teacher retains an awareness of the individual

| | a certain day | the next day |
|---|---|---|
| Topical Inventory | PYTHAGOREAN THEOREM | |
| | History<br>  Euclid's *Elements*<br>Application<br>  (distance btw pts)<br>  Maps, surveying<br>  Carpentering<br>  Sewing<br>Formulation<br>  $a^2 + b^2 = c^2$<br>  square roots<br>  tables, computers | Graphics<br>  geo-board demo<br>  right angle triangle<br>diagonal of square<br>  sides sqrd cells<br>  3, 4, 5, & 5, 12, 13<br>Problem solving<br>  measuring diag's<br>  jaywalkers route<br>Trigonometry |
| **Activity** | **Time spent** | |
| Presentation | xxx | . . . . . . . . . . . . . . xxxx |
| Discussion | xxx | xx |
| Recitation | | |
| Seatwork-drill | xxxxxx | |
| Problem-solving | xxxx | |
| Group work | | xxxxxxxxxxx |
| Notes on individuals | c, d, e, f, g, h,<br>etc. | r, s, t, u, v,<br>etc. |
| | | |
| Main recollection of achievement | "These students did these things" | "These students did these things" |

Figure 5.5.  Teacher recollection of achievement.

student's aptitudes for mathematics, previous experience, and readiness to learn. When required to submit a grade for the term or to brief the students' next teacher about them, the teacher will report such general characteristics. But when assessing what the youngster has *achieved* and how his or her own teaching might be revised, the teacher invokes the inventory of topical experiences and tasks, recalling work completed, noting moments of insight and misunderstanding. My representation of the teacher's perception of student achievement appears as figure 5.5. The emphasis there is on time and tasks, not on student abilities.

On these pages so far, I have described the teaching and learning of mathematics as enormously detailed. It is apparent, on the basis of their words and activities in the classroom, that the conceptualizations of teachers as to what constitutes a course greatly influence their planning, instructional strategy, and assessments. Although the writings of mathematics educators, school district syllabi, textbooks, and tests can be said to be built intellectually on a more powerful structure, these formal conceptualizations of mathematics education do not identify a great many of the characteristics of mathematics achievement important to teachers. Just what information is available from standardized mathematics achievement tests will be dealt with next.

## TESTING AS INFORMATION GATHERING

The public does not understand how there could be sincere objection to using standardized achievement tests to represent what should be taught. People think the tests measure what students know and do not know. They correctly see the tests as nationally based and technically elegant. They presume that tests valid for one educational purpose are valid for others. As students for many years themselves, they have experienced teaching and testing; the question of alignment almost never came up. Today, when a mismatch is claimed, often they presume that the teachers are wrong. Most administrators, counselors, and board members do not question testing. Many interpret objection to testing as an unwillingness to acknowlege shortcomings in the educational system. Dealing with these problems requires a careful look at the standardized achievement test as an information-gathering instrument.

Testing, broadly considered, is the presentation of certain challenges with responses judged as right or wrong. Achievement testing is an activity that generates student performances to be interpreted as above or below pedagogical criteria. A test score indicates performance on a collection of items. Test makers conceptualize what constitutes a good performance. The score is taken as an index of achievement, a datum, a bit of information. For most people, testing is seen mainly as an information-gathering activity. After the information is put to use, we can speak of the validity of the interpretations.

Information from testing can be treated both as measurement data and as data for pondering a problem (Lorge, 1951). When we have measurement data, we think as if we have captured a dimension of something real, something substantial, such as the measurement of age or hat size. For people having measurement appetites, and that includes most of us, standardized tests are expected to provide a trustworthy indicator of student achievement (Cronbach & Meehl, 1955; Shavelson et al., 1987). Even if the scores are not entirely accurate, there is the expectation that the amount of something real is being expressed.

Even when we do not know the validity of the measurements, the information from testing can be problem-pondering data, thought provoking, the basis for hunches, actually helping to shape strategy because strategy is based partly on subjective judgment. These are formative data, potentially useful for redeveloping an idea or practice. It is important for a teacher to consider both possibilities when reviewing test results. Much of achievement testing will fail to provide teachers with dependable measurements, yet be useful for tactical review and reconceptualization; it will provide far more than a guess, far less than a causal link.

Testing is more than information gathering. It is a management control mechanism as well. It is used to announce purpose and priority. It is scheduled in advance by administrators so that effort is shifted toward particular goals. It is an instrument of reward and punishment. Testing and other forms of assessment are widely seen as essential to accountability and educational reform (House, 1973). The effectiveness of the testing process is often seen in terms of the contribution it makes to the management of classroom, school, and school system (Stake, 1991).

In the third part of this chapter, I will present data on the promise, use, and effects of testing—as viewed by a national sample of mathematics teachers. My main interest in the survey was to see what changes, if any, in schoolwork, particularly in curriculum, the teachers attributed to the national emphasis on testing. In the next section, I will discuss the common expectation of testing as an information-gathering activity.

## GENERALIZABILITY OF MATHEMATICS
## KNOWLEDGE FROM STANDARDIZED TESTS

It is widely supposed that a good mathematics test will indicate the amount of student knowledge of the mathematics broadly represented by the test items (Floden et al., 1978). Actually, standardized tests indicate very little as to how much mathematics a student knows (Haertel, 1985; McLean, 1982b). They do not directly measure how well-educated in mathematics the student is becoming. They do not identify the cognitive structures of children's thinking (Easley, 1974; Piaget, 1929).[25] Some indirect inferences can be made by teachers having a good understanding of mathematics as subject matter and how students do mathematics, but the tests add little to what the teacher already knows. Unfortunately, the public is almost invited to make the mistake of concluding that the best learners know all that has been taught and the slowest learners little.

When used with an ordinary group of students, that is, a heterogenous group from the school's catchment area, many standardized achievement tests do effectively indicate which students are the best learners, the ones whose achievement is superior, the ones most quickly becoming educated in the core curriculum of schools. Middle-range and low-performing students also are consistently identified. Tests are poor indicators of future performance of poorly motivated students who subsequently become inspired or of zesty students who in midstream lose interest in academics. But there are few of either of these; approximately the same ranking of students will be found in their courses next year. *Test score interpretation often is valid when predicting standing in the same or an equivalent group at a later time.* The tests can profitably be used to indicate which students probably can handle an accelerated or advanced mathematics course.

With high correlation expected among different mathematics tests, test specialists as well as educators and the general public have come to expect a good test to indicate attainment of knowledge as well as student ranking. With reference to the items of figure 5.1 (shown earlier), these people would expect good performance on Item #1 to indicate a holding of the knowledge needed to do Item #2 and, to a smaller extent, the knowledge needed to do the five less similar items.

It is true that the students who do best on Item #1 will tend to be the ones who do best on Items #2 through #7. But how well a given student will do on those six items is not indicated by performance on the first. Performance on the last item does not indicate how well a student will do on the first six. (*Doing well* here refers to quality of performance on the task regardless of how other students perform.) *Items belonging to a topical family are not necessarily indicators of how well students will do on other items within the family and so, of course, are not indicators of how well students will perform on mathematics items generally.*

For a particular group, Item #1 might be easy. Let us say that 90 percent get it right. Even when the items were selected to have a high interitem correlation, such information about Item #1 gives us no idea about how difficult another item is. Even highly experienced teachers make poor estimates of task difficulty. A group of test items does not provide achievement information on tasks not tested.[26]

Actually, a little more can be deduced. Specialists assembling items for standardized tests want some range of difficulty among items but not too much. To maximize differentiation and to enhance the validity of predictions, they seek items that half the students will get right, half will get wrong. Still, they do not want to discourage examinees by initial items that are too difficult nor to embarrass educators by tests that appear too easy. They try to select a majority of items of about the same difficulty, headed by a few easy items, ending with a few difficult ones. With this knowledge, someone examining the test items, when aware of the difficulty of a few items, can make some guesses as to the difficulty of others. But this serves only as a basis for estimating achievement on other items on the test. It is not a basis for estimating whether or not the examinees would do well on problems not on the test.

As indicated previously, student performance on a great range of mathematics items is remarkably correlated. The more able students tend to find almost all items easier than the less able do. A group of students will show achievement at about the same level on all mathematics items of similar difficulty. Were there to be an inventory of mathematics items having average difficulty equal to the difficulty of items on the test, a group of students would perform about as well en bloc as they performed on the test. It is possible to conceptualize a curricular inventory of mathematics achievement selected solely on the basis of a certain mean difficulty. A teacher noting a mean score on a mathematics test could generalize as to achievement on this generic inventory.

But it would be a fanciful exercise.[27] If we are genuinely interested in the education of youth, we want to see students becoming knowledgeable and skillful as to *particular* mathematics. No domains in mathematics are of absolute value; yet each domain is of more or less value to a robust concept of mathematics education. The inventory, however poorly specified, is where teaching starts. Our lessons, our textbooks, and our tests need to be aligned with what we want mathematics education to be, because they influence what teachers teach and what learners learn. We should not delude ourselves into thinking that domains identified by experts (other than teachers, perhaps) capture the essence of mathematics education. Valid interpretations of achievement scores can reflect different definitions of mathematics education; teacher interpretations will reflect teacher definitions.

Neither specialists in curriculum nor technologists of testing have suitably refined and reported the inventories in *their* heads,[28] much less those in teachers' heads (Archbald & Newmann, 1988). The categories of items of the standardized mathematics achievement test used in the upper sixth grade at Duxbury Intermediate are shown in table 5.3.

The authors go on to identify each item as to number of digits, operation, and units (if any), but it is safe to say that they do not know how to extend the item map into the rest of mathematics. None of us does. None of us commands the language or graphics that details the similarities teachers feel between decimals and fractions, that shows how teachers draw upon understanding of one-step problems to do two-step problems, or that illustrates how

TABLE 5.3.  SKILL CLUSTERS IN THE SRA STANDARDIZED MATHEMATICS ACHIEVEMENT TEST, LEVEL 35, FORM P

| | | |
|---|---|---|
| Computation | Whole Numbers and Money | 17 items; 40 minutes for the section |
| | Decimals | 12 items |
| | Fractions, Mixed Numbers | 10 items |
| Concepts | Whole Numbers and Money | 8 items; 22 minutes for the section |
| | Fractions, Ratios, | 7 items |
| | Proportions & | 6 items |
| | Percents | 4 items |
| | Decimals | 5 items |
| | Prealgebra | |
| | Geometry and Measurement | 7 items |
| Problem Solving | One-Step Problems, | 4 items; 32 minutes for the section |
| | Multiple-Step Problems, Rates, | |
| | Proportion, Percents | 4 items |
| | Geometry Measurement, | 10 items |
| | Statistics | |
| | Problem-Solving Skills | 6 items |

teachers use scale conversions in problem solving. We can give examples, we can demonstrate, we can show how we would teach, but we lack a language to represent those similarities, textures, and relationships. Test producers and others have little linguistic technology for sharing with test users their definitions of mathematics content. The users of mathematics achievement tests are pretty much on their own to decide what mathematics, other than those items actually on the test, is being referred to when we conclude that an examinee is a high achiever.

For a teacher looking closely at the particular items of the test, scores provide a rough indication of how well students would perform subsequently on similar items. Even the best teachers, however, are often wrong in presuming what content is "similar." *Standardized mathematics achievement tests are pretty good at indicating which students are best and which are poorest at learning school mathematics and pretty good at indicating (for those teachers who find it useful to believe in) a general mathematics ability, but quite poor at indicating which knowledge and skill the students have actually attained.*

## ALIGNMENT OF CURRICULUM AND TESTING

Various writers have noted the difference between the curriculum outlined by official goals or syllabi and the curriculum taught by teachers (Eraut, Goad, & Smith, 1975; Berlak et al., 1992; Aoki, 1983). These differences are natural and substantial. Even when coordinators and teachers share both the aims and concepts of teaching method, there is no way for coordinators to state precisely what the teachers should be teaching. Even when the teaching is not very good, the profundity of what goes on in the teaching process defies description. Articulating objectives and teaching in the classroom are two very different media for defining education. In good times, the two provide a dialectic that refines both stating aims and teaching. In bad times, they conflict, they embarrass, they deceive. The differences, of course, are often more than differences in the medium. Official goals and actual practice often project different aims. Values, needs, and conceptualizations can differ, too. In most efforts to reform education, there is a presumption that education would improve if the stated curriculum and the actual curriculum were more congruent.

In the same vein, no textbook perfectly reflects a school's official goals. The outline of content for sixth-grade mathematics

**TABLE 5.4. DISTRICT OUTLINE OF CONTENT AND TEXTBOOK CHAPTER TITLES FOR A SIXTH GRADE MATHEMATICS CLASS**

*District Outline*

1. Addition and Subtraction of Whole Numbers
2. Multiplication and Division of Whole Numbers
3. Introduction to Decimals
4. Multiplication and Division of Decimals.
5. Number Theory
6. Addition and Subtraction of Fractions
7. Multiplication and Division of Fractions
8. Geometry
9. Percent
10. Probability

*Grade Six Textbook Chapters**

1. Addition and Subtraction of Whole Numbers
2. Multiplication and Division of Whole Numbers
3. Decimals
4. Multiplication and Division of Decimals
5. Geometry
6. Factors and Multiples
7. Addition and Subtraction of Fractions
8. Multiplication and Division of Fractions
9. Probability
10. Statistics and Graphing
11. Ratio, Proportion, and Percents
12. Measurement
13. Perimeter, Area, and Volume
14. Integers
15. Using Triangles

*These chapter titles were presented in table 5.1.

in Duxbury in 1989 is compared in table 5.4 to the sixth grade's textbook chapter titles (as seen also in table 5.1). The terms in the two columns match word for word for many topics. The level of detail that district planners and text authors had in mind differs. For example, the Duxbury curriculum guide also lists thirty-three anticipated student outcomes categorized as Knowledge, Skills, and Attitudes. Almost nobody expects a closer match than shown in table 5.4, but it is important to recognize that only the headings match; a great body of detail will not match. An example of a mismatch in teaching fractions would be for the district outline to call for division of sets into subsets as the dominant representation and the textbook to rely almost exclusively on pie charts. A mismatch might be troublesome when the district outline calls for two weeks on geometry and the textbook devotes only five pages to it. It might be a serious mismatch if the textbook only contained exercises on the construction of graphs and district objectives emphasized interpretation of graphs. However, the more the curriculum is in the hands of the teachers, the less we need worry about a mismatch in materials.[29] Usually the teachers take the discrepancy in stride, indifferent to, sometimes too indifferent to, the thrust and boundaries set by both textbook and guide.

And similarly, testing will not and cannot cover precisely the same ground as textbook, syllabus, and especially, teaching practice. In particular teaching practice that follows traditions of local control and teacher autonomy will have different thrusts and boundaries than standardized tests. Here again, differences in language prevent perfect agreement but the differences are likely to be greater than the choices in terms. Test authors and teachers can be expected to differ in their definitions of achievement.

At some level of generality, the curriculum being offered and achievement testing should share an inventory of mathematics content. The word used by researchers (Komoski, no date) in the United States to indicate the match between the inventory to be covered in teaching and the inventory of testing is *alignment*. It seems to most people that teaching and testing should be aligned (Freeman et al., 1983b; Mehrens, 1984; Romberg, 1987, 1992). This is not to say that students should never be tested to determine their understanding of mathematics operations and concepts beyond those taught in the classroom. There are many opportunities to learn mathematics outside the classroom, and many teachers exploit them. When the purpose of testing is primarily to increase understanding of the extent to which a youngster is becoming sophisticated, then the testing inventory should not be limited to school mathematics. *But when the purpose of testing is to increase understanding of the attainment of school mathematics, the curriculum inventory and testing inventory should be aligned.*

Degree of alignment is difficult to measure. As I have already indicated, it is easy to point to examples of misalignment. Some test items will not only be unlike exercises assigned but depend on learning from domains untaught. Some goals, such as "The student will apply the theory and laws governing our number system as known at this level,"[30] are not easily tested. The test, or its extension, the total item pool, is a weak representative of the voluminous inventory of intended learnings. The formal curriculum guide, though usually considerably more detailed than an item pool, is also an understatement of the inventory of desired mathematical achievement. Compared to the curriculum guide, the coverage of the test will fall short. The fact that the test items relate directly to only perhaps 10 percent of the sections in the curriculum guide indicates primarily that the test is much shorter than the guide, not necessarily that there is misalignment between inventories. Matching is problematic. We have no good way of measuring the alignment between tests and curricula.[31]

Still, it is useful for teachers to compare the two. It will become apparent that the test is responsive to some domains and not to others. It may also become apparent that the curriculum guide fails to include all formats in which problems are presented on the test. Usually teachers will decide that students should be expected to solve some problems other than those they actually worked before. A careful review of achievement test items and student performances can broaden and deepen understanding of the complexities of education. Testing can contribute to a provocative consideration of what we mean by mathematics education. On the other hand, a careful review sometimes leads to mindless *teaching to the test*. Alignment usually can be improved by simplifying the curriculum. Whether or not high alignment is good needs to be decided with reference to our basic definitions of education.

My assistant, Giordana Rabitti, studied the alignment among domains in the curriculum guide, textbook, and standardized test for upper sixth grade mathematics in Duxbury. (All of these have been outlined in tables 5.3 and 5.4.) She concluded that a comparison of category headings was insufficient and that a comparison of the details required extensive and highly subjective judgments. From the documents before her, her conclusions could be no more than impressionistic. She found in these materials the potential for a high degree of alignment, but recognized that careful attention by teachers to guide, text, and test would not assure alignment. The curriculum as taught depends on the interpretations of individual teachers who probably will continue to vary widely in their pedagogical methods and inventories of content.

*Test Scores for Redirecting Instruction*

In 1991, as part of President Bush's call for an educational revolution to bring about better schools, it was claimed that reform depended on knowing what each child knows. And *that* knowledge should be obtained with a national student achievement test. An extravagant claim was put forward in 1991 by Secretary of Education Lamar Alexander to the effect that parents have a right to know whether their child understands what is needed as a competitive worker for the world marketplace and what he or she will need to know to be a scientist in the twenty-first century. It is not a matter of parent rights. Only the most primitive knowledge needs of today are known, much less those of tomorrow. Primitive also is our ability to assess what an individual knows, even those who have had extensive research study. The knowledge an ordinary child possesses remains largely unknown. Luckily, effective education systems need only a rough idea of what the learners know. (Those are revolutionary thoughts indeed.)

Teachers have only a rough idea of what children know, yet even this is much more than is to be learned from achievement tests. Standardized tests do not inventory what students know. They tell us mostly which students respond best to the particular questions asked. Standardized test scores represent a reasonable basis for predicting which students will do best in future scholastic assignments but are not a sound basis for redirecting education.

Many people expect standardized achievement tests to have diagnostic properties. State legislation aimed at school reform dramatizes the conviction that the tests will point the road to improvement. Most teachers are skeptical. Over the decades, research studies have made it clear that teachers find standardized test scores of little diagnostic value (Goslin, 1967; Herman & Dorre-Bremme, 1983; Hotvedt, 1974; Tittle, Kelly-Benjamin, & Sacks, 1991). The tests seldom inform teachers of previously unrecognized student talents and seldom identify deficits in a way that directs remedial instruction (Koretz, 1987). Whether referring to an individual student, a class, or a curriculum for the entire country, *standardized achievement tests contribute little to redirecting teaching.*

Is this reflective merely of a shortcoming in our teachers? It could be that teachers are failing to see the diagnostic information embedded in testing and need coaching. Leslie McLean of the Ontario Institute for the Study of Education (1982b) found it more reasonable to conclude that test data do not fit conceptualizations teachers have of education, partly because teacher conceptualizations are more sophisticated. Dale Costello, a British Columbia teacher studying another teacher's classroom, agreed (1988). Teachers increasingly find it useful to look at tests to prepare children to take tests, but not otherwise to redirect their instruction.

Why is remedial action not obvious to a teacher? Suppose that having taken Form P of the SRA test, a group of sixth graders get two-thirds of the computation items correct but two-thirds of the pupils miss the two arithmetic-of-fractions items, #23 and #26. What should the teacher do? Perhaps these items are more difficult; *that* information is not available. Should the teacher allocate additional teaching time to fractions, increasing the risk of not getting to the chapter on probability at all? Is computation more important than activities calling for complex dialogue or engagement in topics of personal interest, neither of which is measured by the test? On the basis of experience, calling on some intuitive sense of the risks involved, each teacher decides. Rational strategies for remediation are missing; heavy emphasis on assessment implies that they should be imposed. *National reform should be based upon teacher conceptualizations of education rather than on student performances on standardized achievement tests.*

We should be wary of premature calls for technology, management by objectives, management by statistical indicators.[32] There may be a technological breakthrough, a massive connection of inventories and pedagogies that relate test performance, immediate instructional tactics, and ultimate educational benefits—someday. Education may someday become a science. So far, both experience and research have provided a few guidelines but not a technology. We have few formal, systematic answers as to what to do about low achievement scores.[33]

Even the informal systems are weak. We might suppose that some educators are so experienced with testing and curriculum development that, given the results of extensive test performances, they can act as consultants to guide curricular change. An Australian researcher, Norman Bowman (1979), spent more than a year in the Midwest, interviewing, observing in many districts, trying to find— wanting very much to find—at least one such expert. He found none. It appears to me that testing people are not interested in curricular epistemology. Curriculum people who want to participate in matters of testing are usually obliged to communicate in the language of behavioral objectives, competencies, and multiple-choice items.[34] Knowledge of the extensive inventories of mathematics, epistemological relationships among domains, and the alternative logics of problem solving are seldom topics for the committee on achievement testing. There is no science of education as a platform for the kind of educational reform President Bush sought.

*Test Score Means Comparing Classes,
Schools, States, and Nations*

With strong backgrounds in the social sciences, where comparisons are major stepping stones for theory construction, test developers produce data for valid comparisons. As noted earlier, their thinking can be categorized in two ways; as either norm referenced or criterion referenced (Glaser, 1963; Hambleton, Algina, & Coulson, 1978).[35] Although these are directly descriptive of an individual examinee, the standardized criterion-referenced tests developed so far are primarily intended for the comparison of individuals.

The comparison of groups has become increasingly common in recent years. It showed up early in program evaluation studies as posttest scores were compared to pretest scores. Later, a public cry for accountability of schools drew attention to the mean score for schools or districts with reference to population norms.[36] Then, researchers raised the question of comparisons among nations, creating the International Education Assessment (MacRury, Nagy, & Traub, 1987) and, in repeated comparisons, found U.S. students performing less well in mathematics than students elsewhere (McKnight et al., 1987). In 1991, comparisons among states became sufficiently political an issue to cause the National Assessment of Educational Progress to pilot comparisons in mathematics (Pipho, 1991).

The standings in these comparisons can be expected to be relatively stable. Low performance of students from the United States, from the southern states, and from urban schools are repeatable and will extend to other domains of scholastic mathematics other than those included in the tests. The problems of interpreting the comparisons are many.[37] According to Harvey Goldstein of the London Institute of Education, "We are still a long way from being able to prescribe a standard analysis which can be adopted routinely to provide definitive school comparions" (1991). Differences in means are much more likely to be attributable to differences in the student groups tested and the alignment of test to teaching than to the quality of teaching. This is not to suggest that teaching quality does not vary widely but that *little can be learned about quality of teaching from student achievement testing.*

The main point to be made about a comparison of group means follows from the point made at the beginning of this section—that test scores can be treated as measurements and as a stimulation to thinking. The comparisons of schools, districts, states, or nations, even though stable, do not measure which mathematics has been learned or how teaching should be redirected. The comparisons can be useful for contemplation: How could performances be this way? How could things be better? Are we putting too much emphasis on geometric proofs? Should we compromise on the use of handheld calculators? The answers to such questions are not to be found in the test scores, but teachers who scrutinize the test scores find new ground for pondering the questions again.


## TESTING AS A MANAGEMENT PROCESS

One of the most powerful realizations in understanding student achievement testing in the schools is the distinction between testing as an information-gathering procedure and testing as an education management procedure. Testing, not just the act itself but the requirement for testing, redirects the teaching process. By requiring standardized testing, politicians and administrators not only make education less a professional field but they change the definition of what an education is.

Traditionally, classroom teaching has been didactic—a matter of exposing the students to certain knowledge, activity, and attitude, then of having them repeat the information, skill, or rule on paper, on a chalkboard, or in recitation (Broudy, 1963). The teacher makes informal assessments of individual progress and decides when to move on to the next topic. For example, according to researcher Ulf Lundgren (1972), it is common for teachers to note particularly the progress of a certain group of students, those consistently at the top of the bottom quarter of the class, in deciding the pace of the teaching. Teacher decisions are moderated, of course, by external demands that certain content, text material, or goals be covered during the year. One way of making those demands is to require standardized testing, particularly "high stakes" testing, in which schooling is placed in jeopardy if test scores are not up to standard.

School reform advocacy in America today is driven by visions of tightened management by central administrators (Eisner, 1992). To improve teaching and learning performance, most citizens and leaders think it necessary to increase commitment to uniform goals and to awaken teachers to their responsibility to the state. If state and district officials are to be accountable to their constituencies, they need to know what is happening—but they have no time to learn. In their view, teaching quality is too complex and obscured to be represented in simple indicators but apparently they believe learning can be so represented, in the form of mean standardized test scores. Thus, school reform appears in public rhetoric to require greater centralization of authority and more powerful information-processing circuits.

In a few places, school reform is happening in just the opposite fashion. In Chicago and New York City, the movement is toward decentralization, toward school-based decision making. In a few places across the country, there are efforts to restructure the schools to draw more upon the professional responsibility of teachers. But such localizing efforts are overwhelmed by the call for core curricula, common goals, and standardized testing. In Sweden and Iceland, objections to control by Stockholm and Reykjavik have resulted in more support for the teacher, less national specification of instruction, and less reliance on standardized testing. But the movement in the United States and most of the world is toward greater control by the government, less honoring of professional experience (particularly as to subject matter conceptualization by the teachers), and more emphasis on formalized student assessment.

Policy analysts such as Linda Darling-Hammond (1991), George Madaus (1991), and Lorrie Shepard (1991) have questioned the plan to orient school reform to national testing. They point to the small payoff from an enormous investment in state-mandated testing. Several recent presidents of the National Council of Measurement in Education, the organization of reference for most educational testing specialists, have expressed strong objection to management expectations of testing mandates (Cole, 1984; Jaeger, 1987, 1992; Madaus, 1985).

Standardized testing in education is an important indicator of authority and control. Teachers who regularly looked first to their experience, then to textbook authors, and last to their professional training to guide their instruction now increasingly appear to look to standardized testing to decide whether their inventories and their conceptualizations of teaching are acceptable (Smith, 1991; Stake, 1991). The degree to which assessment-driven reform is injuring education, or possibly strengthening it, is difficult to assess.

*Perceptions of Teachers as to How Testing Influences Teaching*

There is validity, and there is perception of validity. Invalidity increases as tests are misperceived to be valid and unwarranted interpretations are drawn. When testing is used primarily as a means of exerting pressure for rigorous teaching and learning, resulting actions may be either appropriate or inappropriate. The more that teachers see pressure as appropriate, and many do, the more they are inclined to accept test scores as a valid representation of learning. The more they see mandated testing as inappropriate, the more they stick to their own conceptualizations of education. To understand more of the complexities of test validity, it is important to examine teacher perceptions of the appropriateness of mandated testing and the changes in contemporary schooling they attribute to testing (Hall, Villeme, & Phillippy, 1985; Kemmis et al., 1987; Mattsson, 1989). In this section, I will report results of a national survey I did on mathematics teachers' perceptions of standardized testing in their schools.

The data from this national sample of secondary school mathematics teachers indicate that most teachers, even though they seldom use test information to guide instruction, recognized little invalidity in standardized test information and found the testing little interference to instruction. They found testing still on the increase and associated more with the good changes happening in their schools than the bad. These teacher perceptions may be correct.

When asked to prepare this chapter on validity, I realized that I needed to know more about what was happening in high school mathematics classrooms. I had been observing elementary school classrooms for some time but had not studied mathematics teaching at the high school level since 1978. To help close the gap, I modified a one-page survey form to capture specific concerns raised earlier in this chapter.[38] I retained the survey's basic strategy of asking teachers what changes were happening in their schools, then asking which of those were at least partially attributable to the emphasis on testing.

Colleagues Bernadine Stake and Aminata Soumare and I identified states that had updated lists of mathematics teachers and randomly chose twelve,[39] then made a random selection of high school mathematics teachers from the lists. In the spring of 1990, we sent our single-page questionnaire to one teacher for each 100,000 of state population. We indicated that our purpose was to observe the impact of standardized mathematics achievement tests. We received responses from 186 teachers, a 46 percent return,[40] but had to discard 10 because the respondents were currently not teaching high school mathematics. Soumare did the statistical analysis.

On the whole, the 176 teachers told us that the emphasis on standardized testing in their school was moderate and getting stronger; that is, that it was stronger in 1990 than it had been in 1980, even stronger than it had been in 1987. A third were *supportive* of the increased emphasis but most had *mixed feelings*.[41]

Of the 82 percent whose school had administered a standardized mathematics achievement test within the two previous years, the teachers expressed the following opinions on the validity of the test for representing how well students knew what they had been taught: 45 percent indicated *high validity*; 35 percent *low validity*; and 19 percent responded that they *did not know*. In addition, 58 percent said the test covered content that they had taught; 7 percent said it did not. (It is important to keep in mind the 45 percent indicating *high validity* for representing what they had taught.)

One hundred twenty-five teachers (67 percent) said that over time they had changed their teaching (of the primary topic in their first class of the day they responded). Of those, 78 percent said the change was not because of the testing. Those changing since the last time they taught the course (34 percent) said (again by 79 percent) that the change was not because of the testing. Sharon Dennis, a Seattle mathematics teacher, said she changed "to give more background to effect better total understanding." She found standardized tests to be having *a generally positive effect*.

We asked about the usefulness of testing information to a teacher taking over a class midterm from a teacher who had departed. Only 9 percent said *very important*, 50 percent said *somewhat important*, but 40 percent said *not important at all*. Finally they were asked how often was it useful to review test scores to prepare for a formal conference with parents of a student: 10 percent said *almost always*, 48 percent said *once in a while*, and 40 percent said *almost never*.

About the validity of the test rankings of students on how much mathematics they know: 11 percent (of those responding) found the rankings *highly similar* to their own rankings, 76 percent found *considerable similarity*, and 13 percent *very little similarity*.

We asked how well the standardized tests covered the range of mathematics each was teaching: 4 percent said *extremely well*, 48 percent said *pretty well*, 33 percent said *not very well*, and 5 percent said *extremely poorly*.

As to their opinions about school mean scores as an indication of the quality of teaching provided collectively by the mathematics teachers in the school: 1 percent said they were a *precise indication*, 45 percent said they were a *rough approximation*, and 43 percent said the test scores were a *very poor indication* of teaching quality.

In a summary question, we asked: "Based on what you see in your own school, how is formal standardized testing contributing to efforts to improve the quality of education for the youngsters?" 6 percent said, *it helps in many ways*; 39 percent said, *generally positively; it helps some, but in some ways it hurts*; 17 percent said, *it has no effect*; 16 percent said, *more negatively than positively; helps some but hurts more*; 2 percent said, *it hurts in many ways*.

We noticed differences in the response of teachers in different types of mathematics courses. As indicated in table 5.5, teachers describing themselves primarily as geometry teachers appeared considerably more troubled by the effects of standardized testing than did other mathematics teachers. Teachers whose careers varied in length did not respond differently on the item summarized in table 5.5. Except for the eight teachers who had taught more than thirty years, whose median was a favorable to testing +1.0, the teachers (in five-year experience groups) had medians right at +0.6, the grand median.

The responses summarized in the preceding paragraphs indicate that the teachers found standardized achievement testing in their schools no big problem. They found it adequately aligned with their teaching. About half of them recognized that the tests only partially covered what they were teaching. Most saw some use in the test information.[42] In an earlier survey, Paul Theobald and I (1991) had used many of the same questions with a larger group of elementary school teachers. Their responses were very much the same. During that survey, we talked personally to many of the respondents and found them reluctant to criticize the tests. For

TABLE 5.5. OBSERVATIONS ON THE CONTRIBUTION OF TESTING TO IMPROVING THE QUALITY OF EDUCATION

| Scale | $N =$ | All Teachers 174 | General Math 65 | Algebra Teachers 49 | Geometry Teachers 35 | Advanced Math 64 |
|---|---|---|---|---|---|---|
| 0 | No answer or other answer | 21% | 20% | 22% | 23% | 19% |
| +2 | It helps in many ways | 6% | 8% | 4% | 0% | 5% |
| +1 | Generally positive; helps some, but hurts more | 39% | 37% | 33% | 23% | 39% |
| 0 | It has no effect | 16% | 15% | 20% | 23% | 19% |
| −1 | Negative; helps some, but hurts more | 17% | 20% | 16% | 23% | 17% |
| −2 | It hurts in many ways | 3% | 0% | 4% | 9% | 2% |
| | Median | +0.64 | +0.62 | +0.40 | −0.19 | +0.58 |

example, several said they were grateful that a wayward colleague was forced to bring his or her teaching into line. From many sources (Darling-Hammond, 1991; Shepard, 1991; Smith, 1991), we had heard that teachers were upset by the intrusions of testing on content and schedule, but no more than one out of eight of our respondents voiced vigorous protests against testing. They were uncritical also of the external definition of curriculum and standard setting. They appeared content to teach what outside authorities were telling them to teach.

### Changes in the Schools Attributable to Testing

We asked the same mathematics teachers about changes taking place in their schools, intending later to link some changes to testing. Of the twenty-six we suggested, the following were checked to indicate the *most common changes* in instructional conditions in these schools.[43]

| | |
|---|---|
| We are seeing a gain in emphasis on problem solving and critical thinking. | 59% |
| Advanced courses in high school math are increasingly important here. | 49% |
| Generally there is a broadening of the math curriculum. | 47% |
| There is increasing clarity as to what is to be taught in my math classes. | 46% |
| Teachers are increasingly required to pursue math goals defined by the district. | 45% |
| Attention is increasingly given to differences in individual students. | 45% |
| The marginal learner is increasingly the target for deciding the level to teach at. | 42% |
| Class time spent preparing for tests is increasing. | 40% |
| Our understanding of how much math our students know is increasing. | 40% |
| We are increasing the time we spend on teaching basic math skills. | 39% |

The high school mathematics teachers surveyed indicated that the changes occurring in their schools were what most of us would consider "changes for the good." The most frequently mentioned change was the new emphasis on problem solving and critical thinking. Almost 40 percent noted a continuing increase of emphasis on basic skills. Such an *increase* is a bit hard to comprehend because that emphasis has been strong in most schools for twenty years. It is also a bit hard to interpret the finding that, with all of the emphasis on district goals and curriculum specification, there is increased attention to individual student differences. It might be that more attention is being given to how students score on tests than to the different interests and experiences students have. An increase in class time spent for preparation for standardized testing was one of the most frequently noted changes by these teachers.

We thought it interesting to note also the items checked by these same 176 mathematics teachers as *less frequent changes*:

| | |
|---|---|
| Teachers increasingly utilize those spontaneous "teachable moments." | 34% |
| The quality of our math program is less well understood by administrators. | 30% |
| The quality of our math program is better understood by administrators. | 30% |
| Teachers are increasingly free to pursue the math goals they see important. | 26% |
| Teaching by drawing from the teacher's personal experience is less common. | 26% |
| Teaching by drawing from the teacher's personal experience is more common. | 24% |
| We are diminishing the time we spend on teaching the basic math skills. | 22% |
| Generally there is a narrowing of the math curriculum. | 22% |
| There is increasing confusion as to what is to be taught in my math classes. | 21% |
| Attention is decreasingly given to differences in individual students. | 17% |
| Advanced courses in high school math are becoming less important. | 17% |
| We are seeing a drop in emphasis on problem solving and critical thinking. | 15% |
| The marginal learner is decreasingly the target for deciding the level to teach at. | 14% |

Teachers decreasingly utilize those spontaneous
"teachable moments." 14%
Our understanding of the math our students
know is actually diminishing. 13%
Class time spent preparing for tests is diminishing. 12%

The teachers had numerous opportunities to indicate the negative forces at work in their schools but did not. More exactly, only 10 to 20 percent of the teachers reported such developments. It might be concluded that most teachers saw school reform efforts as working. It could, of course, also have been that the teachers did not see a need for reform. We did not ask them that.

One of the worst fears about testing is that it will narrow the inventory of teaching to those topics and operations included on standardized tests.[44] It is well known that test authors try to avoid content not taught in all schools, thus keeping their tests "fair." Almost 80 percent of the teachers here rejected the opportunity to indicate that the mathematics curriculum was narrowing.

Relatively *stable conditions* (not checked pro or con) were

Utilization of "teachable moments." 52%

Drawing from the teacher's personal experience. 51%

Understanding how much math the students know. 49%

Class time spent preparing for tests. 48%

Targeting teaching on needs of the marginal learner. 44%

These responses suggest that the teachers did not see themselves as losing control of what happens in the classroom. On another matter, half the teachers noted that (even with the stronger districtwide and national emphasis on testing, reported earlier) they are not learning more about how much mathematics the students know. And on still another matter, unlike the 40 percent of teachers who saw increased time being spent preparing for tests, 48 percent of the teachers did not indicate such change.

Now we come to *changes attributed to testing* (by teachers noting that change is at least partly attributable to the school's emphasis on testing):

Teachers are increasingly required to pursue
math goals defined by the district. 80%
We are increasing the time we spend on
teaching basic math skills. 79%

Class time spent preparing for tests is increasing.          73%

Generally there is a narrowing of the
math curriculum.          70%

There is increasing confusion as to what is to
be taught in my math classes.          64%

It is important to recognize that these testing-attribution percentages are based on teachers reporting the change. In the first of the five changes (regarding district goals), 83 of the 176 teachers reported increased requirements; 66 of those 83 (80 percent) attributed the change at least in part to the emphasis on testing. This does not represent a majority of all teachers who claimed testing is having such an effect; it does represent a majority of those reporting increased emphasis on district mathematics goals saying so.

According to a similar minority of teachers, testing is also contributing to the emphasis on basic skills and the narrowing of the curriculum. There is even a perception that testing increases the confusion as to what is to be taught.

The curriculum is constantly changing, and many different social forces and bodies contribute to that change (Freeman et al., 1983b; Saylor, 1982). Often the changes are not planned, but are reactive. Increasing criticism in the news media and the concern of employers about the competence of graduates have caused teachers to narrow the range of learnings for which they are responsible. Seldom are philosophers and educational researchers mentioned as authorities on what education should be. Changes in mathematics curricula occur increasingly in response to public and political expression.

Many people hope that clarity about education will come with increased standardized testing. For more than fifteen years, the states have been mandating achievement testing not so much because it provides useful information but because it accelerates change in educational management (Popkewitz, 1981; Glass & Ellwein, 1986). Testing is supposed to set the classroom teacher straight. I am confident the curriculum has been affected but have found it difficult to measure the amount and direction of change (Stake & Theobald, 1991). Whether standardized test scores are going up or going down has little to tell us about what is happening to education.

During the survey just described, the attitudes of teachers appeared to me blasé. The teachers acknowledged that the inventory of mathematics tested was much smaller than the inventory of mathematics to be taught, yet they did not express concern that these achievement tests provided a limited view of what their students were achieving and not achieving. They seemed to believe that achievement testing was valid if it identified a generalized gradation of mathematics performance among students, thus indicating the "mathematics aptitude" of each student. They had not come to expect the tests to indicate whether or not the students had actually achieved the many curricular components that they, as they taught, treated as separate and important. They were not upset by such test invalidity; they had come to accept it. Therefore, these mathematics teachers, at least the 45 percent who considered the tests as valid and important, were increasingly agreeable to one of the main administrative purposes of testing, to change the views of the teachers as to what should be taught.

## CONCLUSION

We would like our young people, at least in a few areas, to experience the richness and depth of mathematics (de Lange, 1989; Haertel & Calfee, 1983; Romberg, Zarinnia, & Collis, 1989). Instead, as we move toward the challenges of the new century, the American mathematics curriculum has little depth compared to that offered in schools in other industrialized nations (McKnight et al., 1987). Though often well intended, political and technical pressures to specify, standardize, and assess student learning appear to drive the curriculum further toward the shallows. Many mathematics teachers do not recognize the drift toward oversimplification.

It has not helped to upgrade education to specify academic skills and curricular topics as standards for all to master. There are other roads to reform,[45] some that offer increased opportunity for children to *experience* intellectual problems, to voice perplexity, and to propose explanation. Many of us see it as essential that individual children be helped to relate their studies to personalized (*un*common) experience. In trying to raise standards, state school reform efforts have relied excessively on common goals and common test performance.[46] We could not do without common aspirations and expectations, but there is a profound need for unique teaching and allowance for personal interpretation by each child. Many of our teachers are capable of providing it and do. The overemphasis on common goals diverts their efforts.

My field studies of American classrooms (Stake & Easley, 1978; Stake et al., 1986; Stake, Bresler, & Mabry, 1991) indicate that the American teacher remains a major asset, not as capable as we would like, not all that children deserve, but largely pleasing to the local community and school authorities, more the artist and even more the technician than reformist agitation suggests. Most teachers have heard the calls for reform, are sympathetic to them, have helped initiate some, and are hopeful of contributing to improved student assessment. Many are troubled when instructional time is diverted to preparation for testing. Most *do not* see that mandated assessment already is changing the *nature of education* in America.

Education is being redefined. Though the phenomenon is difficult to measure, standardized testing—intentionally, with a noticeable effect that is often harmful—does change education (Haney & Madaus, 1986; Shepard, 1991; Smith, 1991; B. Wilson & Corbett, 1991). As detailed in the previous section, teachers report that with increased testing and curriculum standardization, they attend more to the so-called basics (the most elementary knowledge and skills) and attend less to achieving deep understanding on the part of their students of even a few topics. According to George Madaus (1991), the dangers in current school reform are several: overstandardization, oversimplification, overreliance on statistics, student boredom, increased numbers of dropouts, a sacrifice of personal understanding, and probably, a dimunition of diversity in intellectual development. Further emphasis on standardized testing increases the risk.

Traditionally, education has been a matter of understanding based on knowledge, with each person's knowledge and understanding different because of the impossibility (and undesirability) of completely shared experience. To deal with the complexities of education for both the few and the masses, not only schools but school systems were created. Just as the budget of schools is high on the list of social costs, the management of schools is one of the most comprehensive of collective endeavors. As the authority and accountability of the schools are challenged, school officials look for additional ways of exerting control, not only over learners but over teachers, parents, and taxpayers. The mathematics teachers surveyed for this monograph confirmed that testing is an instrument of management. Those who control the tests have much to say about the definition of learning, teaching, and education (Darling-Hammond & Wise, 1985).

The ostensible purpose of achievement testing is to measure the learning of students. To an extent, this occurs. In measuring mathematics learning, stable rankings among students are obtained. These rankings are not very different from those that teachers have generated informally and more extensively on the basis of classroom observations and assignments. Test scores confirm and authenticate pedagogical assessments. As to what mathematical knowlege has been attained, however, standardized achievement tests provide very little valid information.

Test scores are a stable ground for comparing schools and nations; unfortunately, those comparisons are often distractive, sometimes pernicious. As illustrated in this chapter, they turn teachers and administrators to a lesser task. Comparisons make some deficiencies public but only on the rarest occasion do they show weaknesses not already recognized by teachers and representatives of the public. Seldom do they provide insight or diagnostic remedies for the deficiency. Nor are achievement test scores indicators of the quality of teaching.

To a certain extent, standardized achievement tests should be aligned with the curriculum as planned and taught. The tests should be in harmony with the expectations of parents and the state. Obviously, there is no way to match fully these at least somewhat disparate obligations and expectations. We may someday improve the technology of representing curricular priorities, recognizing with precision what different people want teaching to be, but any reduction in disparity is more likely to be a matter of oversimplying *wants* than of drawing teachers and others into consensus.

An important reason for the lack of alignment between mathematics teaching and standardized achievement testing is the brevity of the tests. Contrary to popular and technical opinion, the items on the test do not nicely represent classroom teachings, even though test items and an abundance of classroom exercises fall within the same goal statement or are classified in the same topical category of mathematics. But the main reason for lack of alignment is that the inventory of mathematics taught by good teachers and poor teachers alike is hundreds of times greater in detail than the mathematics on the test.

Because of the high correlation among student performances on examinations, the test scores provide a stable indicator of some generic notion of mathematics ability. Based on test scores, the predictions that mathematics teachers make about performance in subsequent academic situations will often be valid. The assessment that teachers make (by looking at standardized achievement test scores) about the mathematics that students have learned will seldom be valid. In that sense, there is a fundamental invalidity to standardized testing in mathematics.

APPENDIX

# SURVEY ON SCHOOL TESTING

## Studies of Assessment Policy >> CIRCE, University of Illinois << Mathematics Version

*This survey is to gather actual observations of math teachers on the impact of testing in the schools.*

Thank you for helping.

1. In your own school, what is the present emphasis on standardized testing?   ___ strong.   ___ moderate.   ___ weak.
2. Is that emphasis on testing getting stronger or getting weaker?   ___ getting stronger.   ___ staying the same.   ___ getting weaker.
3. Over all, how do you feel about the emphasis on testing in your school?   ___ opposed to it.   ___ supportive.   ___ mixed feelings.
4. Was the emphasis on testing in your own school greater in 1989 than in 1987?   ___Yes   ___ No   ___I don't know.
5. Was the emphasis on testing in your own school greater in 1989 than in 1980?   ___Yes   ___ No   ___I don't know.

The State Tests. In your own classrooms, different kinds of tests are used. Some are quizzes and examinations authored by a teacher or group of teachers. Some are aptitude tests are not intended to indicate understanding of subject matter but to predict how well students will do in later coursework. Standardized achievement tests are different from all of these. They are tests developed at the state or national level to indicate how well your students have achieved math skills or math-course knowledge. *These standardized math achievement tests are the only tests we are talking about from here on.*

6. Within the last two years or so, has your school administered a standardized math achievement test to students who had enrolled in at least one of your classes?_____.   7. If yes, what was your opinion of the validity of that test for representing how well the students knew what they had been taught? _____ high validity. _____ low validity. _____I don't know.   8. Did that test cover content which you teach? _____.

9. At present, in your first math class of the day, what math topic are you on? _______________________.
10. Have you changed your teaching of this topic since the last time you taught the course? __________.
11. If yes, was that change in part because of the testing? _____.   12. Have you changed your teaching of this topic since you first taught the course? _____.   13. If yes, was the change in part because of state mandate standardized testing? _____.   14. If you did change, briefly describe the change:

15. Suppose somebody in your department leaves suddenly and you have to take over their math class. The students tell you that a standardized math test had been given at the beginning of the course. How important would it be to you to find out how the student had performed on that test?

_____ very important.          _____ somewhat important.          _____ not important at all.

16. When you have a formal conference with parents of your students, how often is it useful to review the scores on standardized math achievement tests?

_____ almost always.          _____ once in a while.          _____ almost never.

17. By the end of a course you teach, you have a good idea of how your students rank among themselves as to how much math they actually know. The standardized math tests used in your school also rank order these students. How similar are these two rank orderings?

_____ highly similar.          _____ considerable similarity.          _____ very little similarity.

18. How well do the standardized math tests used in your school cover the range of math you are teaching?

_____ extremely well.          _____ pretty well.          _____ not very well.          _____ extremely poorly.

19. Are school means on standardized math test a good indication of the quality of teaching provided collectively by the math teachers in your school?

_____ precise indication.          _____ rough approximation.          _____ very poor indication.

Course and level you teach most ___________________________________. Years of math teaching _____.
Your zip code _____. Your name (optional) __________________. Provide address below if you desire feedback. May we quote you by name? _____.

If you've had enough, mail it in. If you can do more, continue on the back

# CHANGING CONDITIONS IN YOUR SCHOOL IN THE LAST YEAR OR TWO

**21** Next we want to identify the statements below which describe what is happening *at your school*. This part is like a True-False test. CIRCLE the number in front of each statement which describes conditions changing in your school during the last year or two. Your circle means the statement describes your school

**23** Now to find connections between the changes occurring in your school and the emphasis on state testing. Please consider each circled statement below, one by one, and ask yourself: "Are these changes caused at least in part by the emphasis on testing?" For each circled item when the answer is "YES" put a check mark (✔) here.

a. Teachers are increasingly required to pursue math goals defined by the district.

b. Teachers are increasingly free to pursue the math goals they see important.

c. Generally there is a narrowing of the math curriculum.

d. Generally there is a broadening of the math curriculum

e. We are increasing the time we spend on teaching basic math skills.

f. We are diminishing the time we spend on teaching basic math skills.

g. Advanced courses in high school mathematics are increasingly important here.

h. Advanced courses in high school mathematics are becoming less important.

i. We are seeing a drop in emphasis on problem solving & critical thinking.

j. We are seeing a gain in emphasis on problem solving & critical thinking.

k. Teaching by drawing from the teacher's personal experience is less common.

l. Teaching by drawing from the teacher's personal experience is more common

m. Class time spent preparing for tests is increasing

n. Class time spent preparing for tests is increasing.

o. The marginal learner is increasingly the target for deciding the level to teach at.

p. The marginal learner is decreasingly the target for deciding the level to teach at.

q.  Teachers increasingly utilize those spontaneous "teachable moments.". . . . . . . .

r.  Teachers decreasingly utilize those spontaneous "teachable moments." . . . . . . . .

s.  There is increasing clarity as to what is to be taught in my math classes. . . . . . . .

t.  There is increasing confusion as to what is to be taught in my math classes. . . . .

u.  Attention is increasingly given to differences in individual students. . . . . . . . . .

v.  Attention is decreasingly given to differences in individual students . . . . . . . . . . .

w.  Our understanding of how much math our students know is increasing. . . . . . . . .

x.  Our understanding of the math our students know is actually diminishing . . . . .

y.  The quality of our math program is better understood by administrators. . . . . . . .

z.  The quality of our math program is less well understood by administrators. . . . . .

You may add additional statements here if you wish

22. OK. The statements you have circled above give us a description of changing conditions in your school. If this description seems biased or misleading, please tell us what it needs to be more correct.

24. OK. By the check marks above, you have told us your perception of the influence of testing on the changes happening in your school. This is very important to us. If you want to emphasize a point please do so here.

25. Now look back at the check marks again. We want to identify the one positive condition in that list of statements that you see as the most important positive contribution testing is making to the changing conditions in your school. Put that number in this box:

26. Now pick the one condition in your school for which testing is making the most negative contribution. Put the number for that statement in this box:

27. And we need a summary statement: Based on what you see in your own school, how is formal standardized testing contributing to efforts to improve the quality of education for the youngsters:

_____ It helps in many different ways.

_____ Generally positively; it helps some, but in some ways it hurts.

_____ It has no effect.

_____ More negatively than positively; helps some but hurts more.

_____ It hurts in many ways.

Thanks for cooperation and thoughtful answers. Bob Stake, Univ. of Illinois (217) 333-3770

## NOTES

1. Extensive survey data will be presented in the third section of this chapter.

2. The current NCTM *Standards* (1989) are part of a long-running campaign by mathematics teacher educators to get teachers to conceptualize less according to topical content and more according to problem solving and experiential learning (Carl, 1991; Biggs & Collis, 1991). The purpose of this chapter is not to argue for one or the other but to examine test validity in terms of teacher and test-developer conceptualizations of mathematics achievement.

3. In speaking of the vast and detailed content that mathematics teachers bring to the classroom, I do not mean to say that as a group they put content learning higher than other learning. Clearly, teachers differ. My own acquaintance with mathematics teachers is nicely reflected in the work of sociologist Robert Connell (1985), who found that teachers tend to prefer one of four emphases: intellectual growth, personal development, skill learning, and honoring custom. Those holding intellect in highest esteem take special pains in choosing content to teach but, whether articulated or not and whether sophisticated or not, all teachers have elaborate conceptualizations of subject matter.

4. Such discourse is at the heart of some definitions of teaching. Speaking of the teacher, Sylvia Ashton-Warner (1967) said: "From the teacher's end it boils down to whether or not she is a good conversationalist; whether or not she has the gift or the wisdom to listen to another; the ability to draw out and preserve that other's line of thought."

5. *Artificial* means that it is a construction of human interpretation and judgment, not a direct inventory or an objectively derived calculation from direct measurements. Mathematics achievement as a construct is artificial because is alludes to a body of material only vaguely specified and largely intuited. Artificial constructs are central to all science. Such constructs as "energy" and "susceptibility to disease" are sometimes objectively defined but are used intuitively and practically by scientists and others. The value of artificial constructs to practitioners depends on how well rooted the concept is in action and discourse.

6. Science, particularly inductive science, is built upon constructs, aggregating through relationships into theories. Testing researchers such as James Popham (1987) and Edward Haertel and David Wiley (1990) speak of domains, traits, and achievements with the confidence that these constructs interchange with the constructs educators develop from experience. The more artificial the construct, the greater is the need for validation, by researcher and educator alike.

7. The emphasis here is on knowledge. Most psychologists prefer to identify the selection of mathematics learned as made up of abilities or competencies. Haertel and Wiley, for example, said, "In this paper the term 'ability' encompasses that which is commonly classified as knowledge and skill" (1990). Such terms draw one toward a concept of education as a

collection of skills and away from thinking of education as understanding of knowledge. Both are part of education but the two definitions move thinking about education in different directions.

8. To appreciate the complexity and lack of interdependence, one might think of an inventory of mathematics the same way one thinks of physical surfaces of land mass. Two-dimensional space could represent knowledge and skill and the elevation could represent conceptual attainment by an individual or group. To each square kilometer, we could assign a learning task. The entire plot might cover a territory as large as a country. For one person, achievement across tasks might be as irregular as the terrain of Switzerland; for another it might be as flat as Holland. Predictions of ground elevation from one part of the country to another would be risky. One cannot indicate ground elevation of a base camp on the Matterhorn from knowledge of evaluation of the railway station in Zurich. And one does not have a very good idea of the elevation of all of Switzerland by sampling elevation at thirty points. Elevation is similar for nearby tasks but attainment of distant tasks is unpredictable.

9. False precision is regularly implied by teachers who grade in percents, implying that 100 percent correct refers to a totality meaningful beyond the items actually administered.

10. Decades ago, most psychometricians abandoned testing for "intelligence" and reconceptualized their target as "scholastic aptitude." Still a form of intellectual power, scholastic aptitude indicates predictable relative achievement on common classroom assignments. Mathematics ability is a specific scholastic aptitude.

11. People will disagree as to who the delinquent teachers are. Shortcomings in subject matter competence, behavior control, punctuality, dress, and test-score production, any one, or more, can be excused by some people when other qualifications run strong. Evaluation of teacher merit is not just a measurement problem; it is confounded by ideological diversity in the school and in the community (Simons & Elliot, 1989; Stiggins & Duke, 1988).

12. Moment by moment, through the day, many students persist in the view that there is little of importance in what the teachers are teaching today. "If it turns out to be important, I can get it later." The indignation of many parents about the schools, often with cause, feeds the youngsters' resistance to being taught.

13. I note also the shortcomings of educational researchers, but to list them in this sentence would imply that they influence what happens in schools.

14. Carl Bereiter (1991) has given us opportunity to rethink the question of how teachers can be effective contributors to student achievement even when they cannot vocalize the rules by which they teach.

15. According to novelist Robertson Davies, the ability to withhold authority and correction is a key talent for teachers. In *The Rebel Angels* (1981, p. 87) he said: "Only those who have never tried it for a week or two can suppose that the pursuit of knowledge does not demand a strength and

determination, a resolve not to be beaten, that is a special kind of energy, and those who lack it or have it only in small store will never be scholars or teachers, because real teaching demands energy as well. To instruct calls for energy, and to remain almost silent, but watchful and helpful, while students instruct themselves, calls for even greater energy. To see someone fall (which will teach him not to fall again) when a word from you would keep him on his feet but ignorant of an important danger, is one of the tasks of the teacher that calls for special energy, because holding in is more demanding than crying out."

16. Research on the automation of education has been summarized by Roy Pea and Elliot Solaway (1987).

17. Swedish researcher Ulf Lundgren (1972) conceptualized the conditions of instruction monitored by teachers. Seymour Sarason (1971) wrote cogently on informal assessment of social conditions in the classroom. John Goodlad wrote about the need for teachers to recognize the quality of student work (1990).

18. Reflective teaching has been described by Donald Schön (1982), by Peter Grimmett and Gaalen Erickson (1988), and by Max van Manen (1991).

19. Some scholars also call for more emphasis on a "core" curriculum. See Fenstermacher and Goodlad (1983).

20. To speak of teaching as an art is not to treat it as casual, contrived, or without standards. Madeleine Grumet (1988, p. 128) notes, "If we think of teaching as an art, then we have a responsibility to be critic as well as artist. To teach as an art would require us to study the transferences we bring to the world we know, to build our pedagogies not only around our feeling for what we know but also around our knowledge of why and how we have come to feel the way we do about what we teach."

21. Speaking of mathematics education in 1967, Robert Davis said, "we live in an age when the best practice of the best practitioners almost certainly lies ahead of the best theory of the best theorists" (p. 59).

22. Figures 5.3, 5.4, and 5.5 are not research findings. They were drawn from my impressions of what experienced teachers do, not directly from observational data. When asked, the teachers seldom claim to be involved in such detailed analyses. And yet the effects of such elaborate selection of content can be observed. The point is that the intuitive working of teachers is highly complex, with far greater texture than the goals stated in table 5.2.

23. The learning terrain for each child, of course, is different.

24. Some of the best works to date are Rosalind Driver, 1973; Bob Godwin, 1990; D. H. Jonassen, 1982; Takahiro Sato, 1991; School Mathematics Study Group, 1961. These works analyze either instruction, epistemology or cognitive development; they do not adapt nicely to the "conversational" exchanges of the American classroom.

25. Other than to note Easley's statement, "it seems absurd to pretend that one knows how to measure cognitive competences by administering standardized lists of questions when no validating clinical interviews—and certainly no explicit structural analyses—have been published" (1974,

p. 281). I will not use this chapter on validity to examine the inability of achievement tests to represent student mathematical thinking. Here I concentrate on the disparity between the inventory of achievement conceptualized by mathematics teachers and the collection of mathematics aptitude items pooled by test makers.

26. Interpreting performance with reference to how other examinees perform, usually using "percentile ranks," is called *norm referencing.* The sophistication of test technology for norm referencing is very high. But, as Robert Glaser (1963), Jason Millman (1974), James Popham (1980), and many educational researchers have said, instruction needs "criterion referencing," interpretation with strong reference to the content of the task. With course content much more rooted in teachers' informal conceptualization than in formal epistemological analysis, the sophistication of criterion-referenced test technology is not very high.

27. Because item difficulties for individual persons are much less stable, this reasoning makes even less sense for interpreting an individual's score.

28. More than in any other subject-matter field, curriculum developers in mathematics have classified content, problems, and skills as to both what might be taught and what should be taught. In their communication, district supervisors and curriculum committees have tended to use only the broad headings (roughly equivalent to textbook chapter titles), classifications that are much less detailed and without the interdependencies that characterize teacher conceptualizations of what needs to be taught (Baker, 1989; Darling-Hammond, 1990).

29. In an unpublished note, Lawrence Stenhouse wrote, "Good teachers are necessarily autonomous in professional judgment. They do not need to be told what to do. They are not professionally the dependents of researchers or superintendents, of innovators or supervisors. This does not mean that they do not welcome access to ideas created by other people at other places or in other times. Nor do they reject advice, consultancy or support. But they do know that ideas and people are not of much real use until they are digested to the point where they are subject to the teacher's own judgment. In short, it is the task of all educationalists outside the classroom to serve the teachers; for only the teachers are in the position to create good teaching."

30. Which was one of the "anticipated student outcomes" for sixth graders in Duxbury.

31. Ken Komoski of the EPIE Institute has been a pioneer in developing an alignment technology.

32. Look at the Vietnam war. The deceit of waging war by strategic use of statistical indicators was illustrated by Neil Sheehan (1988) in his biography of Colonel John Vann. Look at Detroit. According to David Halberstam (1986), the loss of American dominance in the automobile market came about when economists and bankers replaced car makers as chief executive officers. Although spokespersons for business and industry have been strong allies of reform in American schools, some observers of the workplace are urging that worker empowerment, a "real" conceptual

role for workers, increases productivity and corporate health (Peters & Waterman, 1984). Effective teacher-proof management of course content is a pipe dream.

33. One of the finest efforts toward a science of education occurred in the 1970s when Lee Cronbach (1977) and Richard Snow (Snow & Madinach, 1991) of Stanford attempted to pin down relationships among aptitudes (as indicated on tests) and pedagogical strategies. For example, did certain children learn better through practical application whereas others got more from abstract explanations? The findings from that work, unfortunately, warranted little teacher study or technical investment.

34. See P. Tres-Brevig, 1993.

35. A norm-referenced orientation to test development creates a population of examinees whose distribution of scores provide rankings for interpreting each score. The commonly used standardized achievement tests are norm-referenced tests. A criterion-referenced orientation to test development emphasizes individual performances on individual test items selected because, standing alone, that performance is worth pondering. Writing a check a bank would cash, writing an essay, and assembling an engine are such performances.

36. For a close look at the validity of school means, see Madaus et al., 1979. More recently, officials at even the poorest schools found that using the same tests year after year, omitting students with learning deficiencies, and coaching for the tests resulted in district scores above the national median (Linn, Graue, and Sanders, 1990).

37. "Comparisons make sense only when they are put in the context of the entire character of the species concerned and of the known principles governing resemblances between species" (Midgley, 1978, p. 24). For our purposes, substitute the word *schools* or *nations*, for species.

38. The modified instrument and teacher responses are shown in the appendix. Also see Stake & Theobald (1991).

39. Arizona, Connecticut, Idaho, Illinois, Kentucky, Nebraska, New York, South Carolina, Washington, Wisconsin, West Virginia, and Wyoming.

40. Hoping for a much higher return rate, we had kept the questionnaire short and had urged teachers to respond to and send back even a part of it. Having allowed anonymous responses, we could not therefore send follow-up reminders. On the summary item indicated in table 5.5, we calculated the median for those responding in the first three weeks as +0.79, in the second three weeks as +0.46, and for the remainder as +0.56. With these medians, it is reasonable to believe that those who did not respond at all might have been a little more opposed to testing than our respondents were.

41. Response options checked are shown here in italics.

42. Thomas Hastings, Philip Runkel, and Dora Damrin (1961) advised cautiousness in reading teacher's descriptions of test use. Their survey responses indicated a respectable utility, but in probing during personal interviews, they found the testing to be legitimation for decisions more

than important information for decision making. See also the work of Paul LeMahieu (1984) and Romberg, Zarinnia, and Williams (1989).

43. This and the following list consist of a total of twenty-six items *to each of which* teachers could respond or not. Therefore, percentages in the columns will not sum to 100 percent. A third list indicates those issues not checked; that is, considered not to be changing.

44. Teachers have long been dubious about claims for elevating student achievement via stronger control from central administrators. In 1981, the state of Florida had perhaps the most aggressive state testing program in the nation. State Superintendent Turlington repeatedly indicated that Florida teachers were solidly behind the testing program. As a national evaluation team that happened to be studying sex equity education in the nation's tenth largest school district, the Broward County schools, we asked a 15 percent sample of teachers: "In this district's schools, how much are the following interfering with students getting a good education? (a) racial discrimination; (b) discrimination according to sex; (c) bilingual problems; (d) overemphasis on testing." About half the teachers indicated that testing was an interference, more so than racial discrimination, gender discrimination, or bilingualism. (Reported in Stake, Stake, Morgan, & Pearsol, 1983, p. 221.)

45. It is not America alone that is unhappy with its schools. We should not be reluctant to look at how (in the early 1990s) educational authorities around the world are reforming education. Distressed by strict controls from Stockholm, the Swedish people moved Parliament to dismiss the 800-person National Board of Education and replace it with an agency for supporting local educators. The Ministry of Education of Victoria, Australia, has created a system for deciding who shall go to college, using teacher assessment of standardized projects and portfolios. The United Kingdom has piloted "standard assessment tasks," but finds the load on teachers for marking excessive. The Province of Ontario continues to revise its curriculum along lines supported by teacher unions without reliance on state or federal testing. Some ministries seek to draw from science and technology without undermining existing pedagogical arts; others do not.

46. The words of Andrew Porter are instructive: "Simply telling teachers what to do is not likely to have the desired results. Neither is leaving teachers alone to pursue their own predilections. But it might be possible to shift external standard setting away from reliance on rewards and sanctions (power) and toward reliance on authority. One approach to building authoritative standards would be to involve teachers seriously in the business of setting standards. Through the process of teacher participation, the standards would take on authority" (1989, p. 354). The way to involve teachers seriously is to observe what they do rather than ask them what teachers should do. (See also Daniel Koretz, 1987; Ann Lieberman, 1988; National Council of Teachers of Mathematics, 1989; and Harry Torrance, in press.)

## REFERENCES


Airasian, P., & Madaus, G. (1983). "Linking testing and instruction: Policy issues." *Journal of Educational Measurement* 20:103–118.

American Psychological Association, American Educational Research Association, and National Council on Measurement in Education. (1985). *Standards for educational and psychological tests.* Washington, DC: American Psychological Association.

Aoki, T. (1983). "Curriculum implementation as instrumental action and as situational praxis." In T. Aoki (ed.), *Understanding situational meanings of curriculum in-service acts: Implementing, consulting, in-servicing.* p. 98. Alberta: University of Edmonton.

Archbald, D., & Newmann, F. (1988). *Beyond standardized testing: Assessing authentic academic achievement in the secondary school.* Reston, VA: National Association of Secondary School Principals.

Ashton-Warner, S. (1967). *Teacher.* New York: Bantam Books.

Baker, E. (1989). "Mandated tests: Educational reform or quality indicator?" In B. R. Gifford (ed.), *Test policy and test performance: Education, language, and culture.* Boston: Kluwer Academic Publishers.

Bereiter, C. (1991). "Implications of connectionism for thinking about rules." *Educational Researcher* (April):10–16.

Berlak, H., Newmann, F., Adams, E., Archbald, D., Burgess, T., Raven, J., & Romberg, T. (1992). *Toward a new science of educational testing and assessment.* Albany: SUNY Press.

Biggs, J., & Collis, K. (1991). "Multimodal learning and the quality of intelligent behavior." In H. Row (ed.), *Intelligence: Reconceptualization and measurement.* Melbourne: Australian Council on Educational Research.

Bowman, N. (1979). "A search for instances of district use of aggregated test data." Unpublished doctoral dissertation, University of Illinois.

Broudy, H. (1963). "Historic exemplars of teaching method." In N. Gage (ed.), *Handbook of research on teaching.* New York and Chicago: Rand McNally.

Campbell, D., & Fiske, D. (1959). "Convergent and discriminant validation by the multitrait-multimethod matrix." *Psychological Bulletin* 56:81–105.

Carl, I. (1991). "Better mathematics for K–8." *Streamlined Seminar 9,* no. 4 (March):1–7.

Cole, N. (1984). "Testing and the 'crisis' in education." *Educational Measurement: Issues and Practice* 3, no. 3:4–8.

Coley, R., & Goertz, M. (1990). *State educational standards in the fifty states: 1990. ETS RR 90-15.* Princeton, NJ: Educational Testing Service.

Collis, K. (1982). "The SOLO taxonomy as a basis of assessing levels of reasoning in mathematical problem solving." In *Proceedings of the Sixth International Conference for the Psychology of Mathematics Education*, pp. 64–77. Grenoble: University of Grenoble.

———, & Watson, J. M. (1989). "A SOLO mapping procedure." *Proceedings of the Thirteenth International Conference for the Psychology of Mathematics Education*, vol. 1, pp. 180–187. Paris: University of Paris.

Connell, R. (1985). *Teachers' work*. Sydney: George Allen and Unwin.

Costello, D. (1988). "Kaleidoscope patterns: Art education in an elementary classroom." University of British Columbia thesis.

Cronbach, L. (1977). *Aptitudes and instructional methods: A handbook for research on interactions*. San Francisco: Jossey-Bass.

———. (1980). "Validity on parole: How can we go straight? New directions for testing and measurement—Measuring achievement over a decade." In *Proceedings of the 1979 ETS Invitational Conference*, pp. 99–108. San Francisco: Jossey-Bass.

———, & Meehl, P. (1955). "Construct validity in psychological tests." *Psychological Bulletin* 52:281–302.

Darling-Hammond, L. (1990). "Achieving our goals: Superficial or structural reforms." *Phi Delta Kappan* 72, no. 4:286–295.

———. (1991). "The implications of testing policy for educational quality and equality." AERA Conference on Accountability as a State Reform Instrument: Impact on Teaching, Learning, Minority Issues and Incentives for Improvement, Washington, DC.

———, & Wise, A. E. (1985). "Beyond standardization: State standards and school improvement." *Elementary School Journal* 85, no. 3:315–336.

Davies, R. (1981). *The rebel angels*. London: Penguin Books.

Davis, R. (1967). "The range of rhetorics, scale, and other variables." *Journal of Research and Development in Education* 1:51–74.

de Lange, J. (1989). Summary of a presentation to the Mathematical Sciences Education Board. September 22, 1989. University of Utrecht, photocopy.

Department of Employment, Education and Training. (1989). *Discipline review of teacher education in mathematics and science*, vols. 1–3. Canberra: Australian Government Printing Office.

Driver, R. (1973). "The representation of conceptual frameworks in young adolescent science students." University of Illinois dissertation.

Easley, J. (1974). "The structural paradigm in protocol analysis." *Journal of Research in Science Teaching* 2, no. 2:281–290.

———, & Easley, E. (1992). "The curriculum in the classroom: An observational study of mathematics teaching in a Japanese elementary school." In R. Walker (ed.), *Series on classroom research*. Geelong, Australia: Deakin University.

Edelman, M. (1964). *The symbolic uses of politics*. Urbana: University of Illinois Press.

Eiseley, L. (1962). *The mind as nature*. The John Dewey Lectureship, Number Five. New York: Harper & Row.

Eisner, E. (1992). "The federal reform of schools: Looking for the silver bullet." *Phi Delta Kappan*, 73, no. 9:722–723.

Ellwein, M., Glass, G., & Smith, M. L. (1988). "Standards of competence: Propositions on the nature of testing reforms." *Educational Researcher* 17:4–9.

Eraut, M., Goad, L., & Smith, G. (1975). "Analysis of curriculum materials." University of Sussex Education Area Occasional Paper 2, Brighton, England.

Fenstermacher, G., & Goodlad, J. (eds.). (1983). *Individual differences and the common curriculum*. Chicago: University of Chicago Press.

Floden, R., Porter, A., Schmidt, W., & Freeman, D. (1978). *Don't they all measure the same thing? Consequences of selecting standardized tests*. Research Series #25. East Lansing, MI: Institute for Research on Teaching.

Freeman, D., Belli, G., Porter, A., Floden, R., Schmidt, W., & Schwille, J. (1983). "The influence of different styles of textbook use on the instructional validity of standardized tests." *Journal of Educational Measurement* 20, no. 3:259–270.

———, Kuhs, T., Porter, A., Floden, R., Schmidt, W., & Schwille, J. (1983). "Do textbooks and tests define a national curriculum in elementary school mathematics?" *The Elementary School Journal* 83, no. 3:500–513.

Gage, N. (1972). *Teacher effectiveness and teacher education*. Palo Alto, CA: Pacific Books.

Gagné, R. (1967). *Curriculum research and the promotion of learning*, pp. 19–38. AERA Monograph Series on Curriculum Evaluation, 1. Chicago: Rand McNally.

Giroux, H. (1988). *Teachers as intellectuals*. New York: Bergen & Garvey.

Glaser, R. (1963). "Instructional technology and the measurement of learning outcomes." *American Psychologist* 18:519–521.

Glass, G., & Ellwein, M. (1986). "Reform by raising test standards." *Evaluation Comment* (December):1–6.

Godwin, R. (1990). "Epistemic elements in evaluation research." *Studies in Educational Evaluation*, 16:319–333.

Goldstein, H. (1991). *Assessment in schools: An alternative framework*. London: Institute for Public Policy Research.

Goodlad, J. (1990). *Teachers for our nation's schools*. San Francisco: Jossey-Bass.

Goslin, D. (1967). *Teachers and testing.* New York: Russell Sage Foundation.

Greeno, J. (1978). "A study in problem solving." In R. Glaser (ed.), *Advances in instructional psychology* vol. 1. Hillsdale, NJ: Lawrence Erlbaum Associates.

Grimmett, P., & Erickson, G. (1988). *Reflection in teacher education.* Vancouver, BC: Pacific Education Press, University of British Columbia.

Grumet, M. (1988). *Bitter milk: Women and teaching.* Amherst: University of Massachusetts Press.

Haertel, E. (1985). "Construct validity and criterion-referenced testing." *Review of Educational Research* 55, no. 1:23–46.

———, & Calfee, R. (1983). "School achievement: Thinking about what to test." *Journal of Educational Measurement* 20, no. 2:119–132.

———, & Wiley, D. (1990). "Posttest and lattice representations of ability structures: Implications for test theory." Paper delivered at the annual meeting of the American Educational Research Association, Boston.

Halberstam, D. (1986). *The reckoning.* New York: Morrow.

Hall, B., Villeme, M., & Phillippy, S. (1985). "How beginning teachers use test results in critical decisions." *Educational Research Quarterly* 9, no. 3:12–18.

Hambleton, R. (1989). "Principles and selected applications of item response theory." *Educational Measurement* 3:147–200.

———, Algina, J., & Coulson, D. B. (1978). "Criterion referenced testing and measurement: A review of technical issues and developments." *Review of Educational Research* 48:1–47.

Haney, W., & Madaus, G. (1986). "Effects of standardized testing and the future of the national assessment of educational progress." Working paper for the NAEP. Chestnut Hill, MA: Study Group on the National Assessment of Student Achievement.

Hastings, T., Runkel, P., & Damrin, D. (1961). *Effects on use of tests by teachers trained in a summer institute.* Urbana: University of Illinois, Bureau of Educational Research.

Henderson, K. (1963). "Research on teaching secondary school mathematics." In N. Gage (ed.), *Handbook of research on teaching.* Chicago: Rand McNally.

Herman, J., & Dorre-Bremme, D. (1983). "Uses of testing in the schools: A national profile." *New Directions for Testing and Measurement* 19.

Hively, W. II, Maxwell, G., Rabehl, G., Sension, D., & Lundin, S. (1973). *Domain-referenced curriculum evaluation: A technical handbook and a case study from the Minnemast Project.* CSE Monograph Series in Evaluation 1. Los Angeles: Center for the Study of Evaluation, UCLA.

Hotvedt, M. (1974). "A case study of standardized test use in public school." Unpublished doctoral dissertation, University of Illinois, Urbana.

House, E. R. (ed.). (1973). *School evaluation: The politics and process.* Berkeley, CA: McCutchan.

Jaeger, R. (1987). "Two decades of review of educational measurement!" *Educational Measurement: Issues and Practices* 6:6–14.

———. (1992). "World class standards, choice, and privatization: Weak measurement serving presumptive policy." Vice-presidential address, American Educational Research Association, San Francisco.

———, & Tittle, C. (eds.). (1980). *Minimum competency achievement testing: Motives, models, measures, and consequences.* Berkeley, CA: McCutchan.

Johnson, D., & Johnson, R. (1991). *Learning together and alone,* 3rd ed. Englewood Cliffs, NJ: Prentice-Hall.

Jonassen, D. H. (ed.). (1982). *The technology of text.* Englewood Cliffs, NJ: Educational Technology Publications.

Kemmis, S., McTaggart, R., Smyth, J., & Ross, K. (1987). "Monitoring school performance: Testing and alternatives to it." Geelong, Australia: Deakin University. Photocopy.

Komoski, K. (No date). "Curriculum Analysis Report. (Using the Integrated Instructional Information Resource Data Base in the Battle Creek Public Schools)." Watermill, NY: EPIE Institute.

Koretz, D. M. (1987). *Educational achievement: Explanations and implications of recent trends.* Washington, DC: U.S. Congressional Budget Office.

LeMahieu, P. (1984). "The effects of achievement and instructional content on a program of student monitoring through frequent testing." *Educational Evaluation and Policy Analysis* 6, no. 2:175–187.

Lampert, M. (1988). *The teacher's role in reinventing the meaning of mathematical knowing in the classroom.* Research Series 186. Lansing: Michigan State University, Institute for Research on Teaching.

Lieberman, A. (1984). *Teachers, their world and their work.* Alexandria, VA: Association for Supervision and Curriculum Development.

———. (1988). *Building a professional culture in schools.* New York: Teachers College Press, Columbia University.

Linn, R. (ed.). (1989). *Educational measurement,* 3rd ed. New York: Macmillan and American Council on Education.

———, Graue, E., & Sanders, N. (1990). "Comparing state and district test results to national norms: The validity of the claim that 'everyone is above average.' " *Educational Measurement: Issues and Practice*: 5–14.

Lorge, I. (1951). "The fundamental nature of measurement." In E. F. Lindquist (ed.), *Educational measurement*. Washington, DC: American Council on Education.

Lortie, D. (1975). *School teacher*. Chicago: University of Chicago Press.

Lundgren, U. (1972). *Frame factors and the teaching process*. Stockholm: Almqvist & Wiksell.

MacRury, K., Nagy, P., & Traub, R. (1987). *Reflections on large-scale assessments of study achievement*. Toronto: Ontario Institute for Studies in Education.

Madaus, G. (1985). "Test scores as administrative mechanisms in educational policy." *Phi Delta Kappan* 66:611–617.

———. (1991). "The effects of important tests on students: Implications for a national examination or system of examinations." AERA Conference on Accountability as a State Reform Instrument: Impact on Teaching, Learning, Minority Issues and Incentives for Improvement. Washington, DC.

———, Kellaghan, T., Rakow, E. A., & King, D. J. (1979). "The sensitivity of measures of school effectiveness." *Harvard Education Review* 49, no. 2:207–229.

Marvin, C. (1988). "Attributes of authority: Literacy tests and the logic of strategic conduct." *Communication* 11:63–82.

Mattsson, H. (1989). *Tests in school—seen through the eyes of teachers*. Umeå, Sweden: Umeå University.

McKnight, C., Travers, K., & Dossey, J. (1985). "Twelfth-grade mathematics in U.S. high schools: A report from the second international mathematics study." *Mathematics Teacher* 78, no. 4:292–300.

McKnight, C., Crosswhite, J., Dossey, J., Kifer, E., Swafford, J., Travers, T., & Cooney, T. (1987). *The underachieving curriculum: Assessing U. S. school mathematics from an international perspective*. Champaign, IL: Stipes Publishing.

McLean, L. (1982a). *Report of the 1981 field trials in English and mathematics: Intermediate division*. Toronto: The Minister of Education.

———. (1982b). "Achievement testing—Yes! Achievement tests—No." *E+M Newsletter* 39:1–2.

Mehrens, W. (1984). "National tests and local curriculum: Match or mismatch?" *Educational Measurement: Issues and Practice* 3, no. 3:9–15.

Messick, S. (1989). "Validity." In R. Linn (ed.), *Educational measurement*, 3rd ed., pp. 13–103. New York: Macmillan and American Council on Education.

Midgley, M. (1978). *Beast or man*. New York: Meridian Books.

Millman, J. (1974). "Criterion referenced measurement." In W. J. Popham (ed.), *Evaluation in education*, pp. 309–398. Berkeley, CA:McCutchan.

National Council of Teachers of Mathematics (NCTM). (1989). *Curriculum and evaluation standards for school mathematics*. Reston, VA: Author.

Oakes, J. (1985). *Keeping track: How schools structure inequality*. New Haven, CT: Yale University Press.

Oliver, D. (1988). *Education, modernity, and fractured meaning*. Albany: State University of New York Press.

Pea, R. (1987). "Socializing the knowledge transfer problem." *International Journal of Educational Research* 11, no. 6:639–663.

———, & Soloway, E. (1987). *Mechanisms for facilitating a vital and dynamic educational system: Fundamental roles for educational science and technology*. Final report for the U.S. Office of Technology Assessment. New Haven, CT: Cognitive Systems.

Peters, T., & Waterman, R., Jr. (1984). *In search of excellence*. New York: Warner Books.

Piaget, J. (1929). *The child's conception of the world*. London: Routledge and Kegan Paul.

Pipho, C. (1991). "The unbridled, undebated national test." *Phi Delta Kappan* 72, no. 8:574–575.

Popham, J. (1980). "Domain specification strategies." In R. Berk (ed.), *Criterion-referenced measurement: The state of the art*. Baltimore: Johns Hopkins University Press.

———. (1987). "The merits of measurement-driven instruction." *Phi Delta Kappan* 68, no. 9:679–682.

Popkewitz, T. (1981). *The myth of educational reform: A study of school responses to a program of change*. Madison: University of Wisconsin Press.

Porter, A. (1989). "External standards and good teaching: The pros and cons of telling teachers what to do." *Educational Evaluation and Policy Analysis* 11, no. 4:354.

Proppé, O., Myrdal, S., & Danielsson, B. (In preparation). "Teacher education reforms: A seven-country study of changing patterns of regulations." Reykjavik: The Icelandic University College of Education.

Raven, J. (1992). "A model of competence, motivation, and behavior, and a paradigm for assessment." In H. Berlak, F. Newmann, E. Adams, D. Archbald, T. Burgess, J. Raven, & T. Romberg (eds.), *Toward a new science of educational testing and assessment*, pp. 85–116. New York: State University of New York Press.

Resnick, D. (1981). "Testing in America: A supportive environment." *Phi Delta Kappan* 62, no. 9:625–628.

Resnick, L. (1989). "Developing mathematical knowledge." *American Psychologist* 44, no. 2:162–169.

Rice, M., & Mouseley, J. (No date). "Creativity and social reproduction in mathematics curriculum." Geelong, Australia: Deakin University, photocopy.

Robitaille, D. (ed.). (1989). *Evaluation and assessment in mathematics education*. Paris: UNESCO.

Romberg, T. A. (1983). "A common curriculum for mathematics." In G. D. Fenstermacher & J. Goodlad (eds.), *Individual differences and the common curriculum*. Chicago: National Society for the Study of Education and University of Chicago Press.

———. (1985). "The content validity of the mathematics subscores and items for the second international mathematics study." Paper presented to the Committee on National Statistics, National Research Council of the National Academy of Sciences. Madison: Wisconsin Center for Education Research.

———. (1987). *The domain knowledge strategy for mathematical assessment*. Madison, WI: National Center for Research in Mathematical Sciences Education.

———. (1992). "Assessing mathematics competence and achievement." In H. Berlak, Newmann, F., Adams, E., Archbald, D., Burgess, T., Raven, J., & Romberg, T. (eds.), *Toward a new science of educational testing and assessment*. Albany: SUNY Press.

———, & Carpenter, T. P. (1986). "Research on teaching and learning mathematics: Two disciplines of scientific inquiry." In M. Wittrock (ed.), *Handbook of research on teaching*. New York: Macmillan.

———, Zarinnia, A., & Collis, K. (1989). In G. Kulm (ed.), *Assessing higher order thinking in mathematics*. Washington, DC: American Association for the Advancement of Science.

———, Zarinnia, A., & Williams, S. R. (1989). *The influence of mandated testing on mathematics instruction: Grade 8 teachers' perceptions*. Madison, WI: National Center for Research in Mathematical Sciences Education.

Rosenshine, B. (1970). *Teaching behaviors and student achievement*. Stockholm: International Association for the Evaluation of Educational Achievement.

Sarason, S. (1971). *The culture of the school and the problem of change*. Boston: Allyn & Bacon.

Sato, T. (1991). *Development of instructional materials in NEC Technical College: Concepts and practices*. Tokyo: NEC Corporation.

Saylor, G. (1982). *Who planned the curriculum?* West Lafayette, IN: Kappa Delta Pi.

Scheffler, I. (1965). *Conditions of knowledge*. Chicago: Scott, Foresman and Company.

———. (1973). *Reason and teaching*. Indianapolis: Bobbs-Merrill.

———. (1975). "Basic mathematical skills: Some philosophical and practical remarks." In *The NIE Conference on Basic Mathematical Skills and Learning*, vol. 1, pp. 182–189. Los Alamitos: Southwest Regional Laboratory.

School Mathematics Study Group. (1961). *First course in algebra: Units 9 and 10*. New Haven, CT: Yale University Press.

Schön, D. (1982). *The reflective practitioner: How professionals think in action*. New York: Basic Books.

Shavelson, R., McDonnell, L., Oakes, J., & Carey, N. (1987). *Indicator systems for monitoring mathematics and science education*. Santa Monica, CA: Rand Corporation.

Sheehan, N. (1988). *A bright and shining lie: John Vann and America in Vietnam.* New York: Random House.

Shepard, L. (1991). "Will national tests improve student learning?" AERA Conference on Accountability as a State Reform Instrument: Impact on Teaching, Learning, Minority Issues and Incentives for Improvement, Washington, DC.

Shohamy, E. (1984). "Does the testing method make a difference? The case of reading comprehension." *Language Testing* 1, no. 2:147–170.

Simons, H., & Elliott, J. (eds.). (1989). *Rethinking appraisal and assessment*. Philadelphia: Open University Press.

Smith, M. L. (1991). "Put to the test: The effects of external testing on teachers." *Educational Researcher*, 20, no. 5:8–11.

Snow, R., & Lowman, D. F. (1989). "Implications of cognitive psychology for educational measurement." In R. Linn (ed.), *Educational Measurement*, New York: Macmillan and American Council on Education. 3rd. ed., pp. 263–331.

Snow, R., & Mandinach, E. (1991). *Integrating assessment and instruction: A research and development agenda*. RR-91-8. Princeton, NJ: Educational Testing Service.

Stake, B., Stake, R., Morgan, L., & Pearsol, J. (1983). *Evaluation of the national sex equity demonstration project final report, 1980–1983*. Urbana: Center for Instructional Research and Curriculum Evaluation, University of Illinois.

Stake, R. (ed.). (1991). *Using assessment policy to reform education*. Advances in Program Evaluation, 1A. Greenwich, CT: JAI Press.

———, Bresler, L., & Mabry, L. (1991). *Custom and cherishing: The arts in elementary schools*. Urbana: University of Illinois, Council for Research in Music. Music Educators National Conference.

———, & Easley, J. (eds.). (1978). *Case studies in science education*. Urbana: Center for Instructional Research and Curriculum Evaluation, University of Illinois.

———, Raths, J., Denny, T., Stenzel, N., & Hoke, G. (1986). *Evaluation study of the Indiana department of education gifted and talented program*. Urbana: Center for Instructional Research and Curriculum Evaluation, University of Illinois.

———, & Theobald, P. (1991). "Teachers' views of testing's impact on classrooms." *Advances in Program Evaluation*, 1B:189–201.

Stiggins, R., & Duke, D. (1988). *The case for commitment to teacher growth.* Albany, NY: SUNY Press.

Tittle, C., Kelly-Benjamin, K., & Sacks, J. (1991). "The construction of validity and effects of large-scale assessments in the schools." *Advances in Program Evaluation* 1B:233–254.

Torrance, H. (In press). "Evaluating SATs—the 1990 pilot." *Cambridge Journal of Education.*

Traxler, A. (1951). "Administering and scoring the objective test." In E. F. Lindquist (ed.), *Educational measurement.* Washington, DC: American Council on Education.

Tres-Brevig, M. de P. (1993). *Effects of implementation of assessment policy on staff practices at a state department of education.* University of Illinois Ph.D. dissertation in draft.

United Nations Educational, Scientific, and Cultural Organization. (1989). Qualities required of education today to meet foreseeable demands in the twenty-first century. Symposium, Beijing, 29 November–2 December, 1989.

van Manen, M. (1991). *The tact of teaching.* Albany: SUNY Press.

Vannatta, G., & Stoeckinger, J. (1980). *Mathematics: Essentials and applications.* Columbus, OH: Charles E. Merrill Publishing Company.

Willis, S. (1990). *Science and mathematics in the formative years.* Australian Government Printing Service.

Wilson, B., & Corbett, D. (1991). "Two state minimum competency testing programs and their effects on curriculum and instruction." *Advances in Program Evaluation* 1B:pp. 7–40.

Wilson, J. (1971). "Evaluation of learning in secondary school mathematics." In B. Bloom, T. Hastings, & G. Madaus (eds.), *Handbook on formative and summative evaluation of student learning*, pp. 645–696. New York: McGraw-Hill.

# 6 ❖ Assessment Nets: An Alternative Approach to Assessment in Mathematics Achievement

*Mark Wilson*

New views of learning have implications for the monitoring and assessment of student learning (Webb & Romberg, 1992). They suggest that we focus on measuring the understanding and models that individual students construct during their learning process (Masters & Mislevy, 1992; Wilson, 1992b). In many areas of learning, and in mathematics in particular, levels of achievement may best be defined and measured not in terms of the *number* of facts and procedures that a student can reproduce (i.e., test score as counts of correct items), but in terms of the best estimates of his or her *level* of understanding of its key concepts and principles (Masters, Adams, & Wilson, 1990; Wolf, Bixby, Glenn, & Gardner, 1991). Moreover, we may need to estimate these levels using several types of information from a single complex performance (e.g., levels of correctness, strategy-use, latency), and we may need to incorporate several perspectives on the types of information (e.g., the student's perspective, the teacher's perspective, an expert opinion). Obtaining these types of information and perspectives on the information will require new approaches to assessment.

This chapter describes an *assessment net* that is an alternative form of assessment for mathematics education. An assessment net is composed of (1) a *framework* for describing and reporting the level of student performance, (2) a means of *gathering information* based on observational practices that are consistent with both the educational variables to be measured and the context in which that measurement is to take place, and (3) a measurement model that provides for appropriate forms of *quality control.*

To date, work in the area of performance assessment has addressed only one portion of an "assessment system" (Linn, Baker, & Dunbar, 1991), observational design, and emphasized instructional validity (Wolf, Bixby, Glenn, & Gardner, 1991). For example, a recent issue of a journal concerned with measurement in education was devoted to performance assessment (Stiggins &

236

Plake, 1991), yet only one article dealt substantively with issues other than information gathering.

At this time we do not have a comprehensive methodology for performance assessment. The complexity of performance assessment appears to challenge both the philosophical foundations (Shepard, 1991) and the technology (i.e., the measurement models) of standard educational and psychological measurement. In contrast, the rival of performance assessment, standardized multiple-choice testing, appears to many to be part of a coherent system of assessment (APA, AERA, & NCME, 1985) that ensures quality control by addressing item test construction, pilot testing, reliability, validity, and reporting schemes. The aim of the assessment net is to build a system that has the coherence of the traditional testing approaches but addresses new issues brought forward by the performance assessment movement.

### FRAMEWORK

The assessment net begins with the idea that what we want to assess is student *progression* along the strands of the curriculum. This progression must reflect a shared understanding on the part of the users of the net. That understanding must include a notion of progression, an agreed-upon set of important strands, an agreed-upon set of levels of performance along the strands, and an acceptance that this progression is a tendency but not an absolute rule. A *framework* for a particular curriculum area defines levels of performance that students would be expected to achieve. The levels extend from lower, more elementary knowledge, understanding, and skills to more advanced ones. They describe understanding in terms of qualitatively distinguishable performances along the strands or continuum.

The idea of a framework is not new. Related notions have been developed in many parts of the world: the Western Australia *First Steps* project (Ministry of Education, 1991), the Australia National Curriculum Profiles (Australia Education Council, 1992), and the UK National Curriculum strands (Department of Education and Science, 1987a, 1987b). The California Framework in mathematics (California State Department of Education, 1985) is composed of strands or continua in number, measurement, geometry, patterns and functions, statistics and probability, and algebra. Within each of these strands, four broad levels of performance are defined: (1) kindergarten to grade 3, (2) grades 3 to 6, (3) grades 6 to 8, and

(4) grades 9 to 12. A list of goals is defined within each strand for each level. For example, within the geometry strand, at the lowest level, one of the goals is "Use visual attributes and concrete materials to identify, classify, and describe common geometric figures and models, such as rectangles, squares, triangles, circles, cubes, and spheres. Use correct vocabulary" (California State Department of Education, 1985, p. 24). At the grades 3–6 level, one of the goals is "Use protractor, compass, and straightedge to draw and measure angles and for other constructions" (California State Department of Education, 1985, p. 27). And at the grades 6–8 level, one of the goals is "Describe relationships between figures (congruent, similar) and perform transformations (rotations, reflections, translations, and dilations)" (California State Department of Education, 1985, p. 32). Each level is associated with a set of special concerns and emphases for that particular period of schooling, such as, at the base level, an emphasis on concrete materials and classification.

The Vermont statewide assessment for fourth and eighth grade students includes standard tests and portfolios in mathematics and writing. Students' mathematics portfolios are rated on the seven criteria shown in figure 6.1: understanding of task, quality of approaches and procedures, decisions along the way, outcomes of activities, language of mathematics, mathematical representations, and clarity of presentation (Vermont Department of Education, 1991).

The mathematics projects used in statewide assessment in Victoria, Australia, are a third example. Students in the final two years of high school undertake studies for a certificate issued by the Victoria Curriculum and Assessment Board. To satisfy the requirements for the certificate, students complete twenty-four half-year units chosen from forty-four available areas of study. All students must complete a specified number of units in English, arts-humanities, and mathematics-science-technology.

Students who take mathematics are required to complete a series of investigative projects, each involving at least seven hours of classwork. One of these projects, completed during the first half of the final year, is based on a theme set annually by the board. Teachers monitor and record the progress of each student's project, much of which must be completed during class time. The project report is submitted by a date specified by the board. The task is described as follows: "Students undertake an independent mathematical investigation based on a single theme set annually by the state curriculum and assessment board. Students have four weeks

1. <u>Understanding of Task</u>

*Sources of Evidence:* Explanation of task; reasonableness of approach; correctness of response leading to inference of understanding.

    3 Generalized, applied, extended.
    2 Understood.
    1 Partially Understood.
    0 Totally Misunderstood.

2. <u>Quality of Approaches/Procedures</u>

*Sources of Evidence:* Demonstrations: descriptions (oral or written); drafts, scratch work, etc.

    3 Efficient or sophisticated approach/procedure.
    2 Workable approach/procedure.
    1 Appropriate approach/procedure some of the time.
    0 Inappropriate or unworkable approach/procedure.

3. <u>Decisions Along the Way</u>

*Sources of Evidence:* Changes in approach; explanations (oral or written); validation of final solution; demonstration.

    3 Reasoned decisions/adjustments shown/explicated.
    2 Reasoned decisions/adjustments inferred with certainty.
    1 Reasoned decision making possible.
    0 No evidence of reasoned decision making.

4. <u>Outcomes of Activities</u>

*Sources of Evidence:* Solutions; extensions—observations, connections, applications, syntheses, generalizations, abstractions.

    3 Solution with synthesis, generalization, or abstraction.
    2 Solution with connections or application(s).
    1 Solution with observations.
    0 Solution with extensions.

5. <u>Language of Mathematics</u>

*Sources of Evidence:* Terminology; notation/symbols.

    3 Use of rich, precise, elegant, appropriate mathematical language.
    2 Appropriate use of mathematical language most of the time.
    1 Appropriate use of mathematical language some of the time.
    0 No or inappropriate use of mathematical language.

6. <u>Mathematical Representations</u>

*Sources of Evidence:* Graphs, tables, charts; models; diagrams; manipulatives.

    3 Perceptive use of mathematical representation(s).
    2 Accurate and appropriate use of mathematical representation(s).
    1 Use of mathematical representation(s).
    0 No use of mathematical representation(s).

7. <u>Clarity of Presentation</u>

*Sources of Evidence:* Audio/video tapes (transcripts); written work; teacher interviews/observations; journal entries; student comments on cover sheet; student self-assessment.

    3 Clear (e.g., well-organized, complete, detailed).
    2 Mostly clear.
    1 Some clear parts.
    0 Unclear (e.g., disorganized, incomplete, lacking detail).

Figure 6.1. Criteria for mathematics portfolio.

to develop a project topic based on the theme, collect data, and submit a written report. The project is expected to take between 15 and 20 hours, with 7 to 10 hours during class time. Each student submits a written report of about 1,500 words emphasizing the mathematical aspects and results of the project. This task is undertaken in the middle of the school year. An initial grade is assigned by the school and assessments are subject to the verification procedures of the Board" (Victoria Curriculum and Assessment Board, 1990).

Initial assessments of students' projects are completed by classroom teachers. To achieve comparability for the award of project grades across schools throughout the state, all teachers are provided with an assessment sheet listing a set of eighteen criteria:

(a) Conducting the investigation
    1. Identifying important information
    2. Collecting appropriate information
    3. Analyzing information
    4. Interpreting and critically evaluating results
    5. Working logically
    6. Broadening or deepening the investigation
(b) Mathematical content
    7. Mathematical formulation or interpretation of problem situation or issue
    8. Relevance of mathematics used
    9. Level of mathematics used
    10. Mathematical language, symbols, and conventions used
    11. Understanding, interpreting, and evaluating the mathematics used
    12. Accuracy of mathematics used
(c) Communication
    13. Clarity of aims of project
    14. Relation of project topic to theme
    15. Definitions of mathematical symbols used
    16. Account of investigation and conclusions
    17. Evaluation of conclusions
    18. Organization of material (Victoria Curriculum and Assessment Board, 1990)

Teachers rate each student's project as *high, medium, low,* or *not shown* on each of these eighteen criteria. The concrete meaning of the levels comes from the way teachers observe students and rate them.

A second class of assessment involves focused applications that are based on particular pedagogic or developmental theories. These typically have definitions of levels that are not arbitrary; rather, they are based on a theory. Two such examples follow, one from a psychological research setting, and one from educational research.

Siegler (1987) has described a study in which students were presented with a series of elementary addition problems and then were asked, "How did you figure out the answer to that problem?" Their answers were classified into one of five categories according to a scheme based on earlier research:

1. Retrieval (R), where the student retrieves the answer from memory.
2. Min strategy (M), where the student counts up from the larger addend the number of times indicated by the smaller addend.
3. Decomposition (D), where the student transforms the original problem into two or more simpler problems.
4. Counting-all strategy (C), where the student counts from one the number of times indicated by the sum.
5. Guessing and "other" (G), where the student says that he or she guessed or did not know the answer.

In his paper, Siegler makes the points that (a) dependent variables such as solution time and error rate should not be "averaged over" these strategies as they were in the past, an approach that led to contradictory results in studies of addition, and (b) students do not use one strategy exclusively, but tend to show substantial variation. His analyses show clearly that some strategies are "better" than others in that they are associated with speed or a lower error rate. The literature provides evidence of a developmental sequence among the addition strategies. Ashcraft (1982) found that, although first graders are fairly consistent in their use of the Min strategy, fourth graders consistently use Retrieval, and third graders use a mixture of the two strategies. Siegler challenged the interpretation of Ashcraft's results, but he also found that frequency of use of strategies changes according to grade level. Although many questions about the use of strategies remain, most are best answered by considering strategies one at a time. Perhaps strategies will be seen as part of a strategy-use continuum that can summarize not only the development of strategies, but also any regularities in their distribution within and between individuals.

Figure 6.2 shows the levels of strategy-use for addition problems that are supported in empirical research (Wilson, 1992c). Note that the ordering of the strategies is not linear—different strategies may be equally "good" at a developmental level.

A second example uses phenomenography. Phenomenographic analysis has its origins in the work of Marton (1981), who describes it as "a research method for mapping the qualitatively different ways in which people experience, conceptualize, perceive, and understand various aspects of, and phenomena in, the world around them" (Marton, 1986, 31). Phenomenographic analysis usually involves the presentation of an open-ended task, question, or problem designed to elicit information about an individual's understanding of a particular phenomenon. Often tasks are attempted in relatively unstructured interviews during which students are encouraged to explain their approach to the task or their conception of the problem. Researchers have applied phenomenographic analysis to such learning areas as proportionality (Lybeck, 1981), number (Neuman, 1987), and speed, distance, and time (Ramsden, 1990).

These studies found that students' responses reflect a limited number of qualitatively different ways of thinking about a phenomenon, concept, or principle (Marton, 1988). These outcome

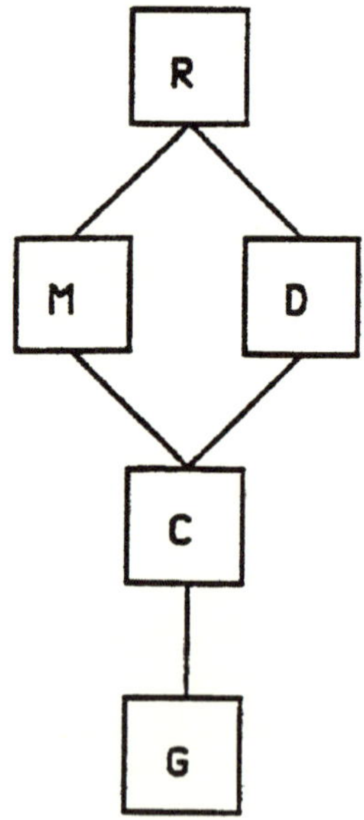


Figure 6.2. Strategy-use levels.

categories are "usually presented in terms of some hierarchy: There is a *best* conception, and sometimes the other conceptions can be ordered along an evaluative dimension" (Marton, 1988, p. 195). For Ramsden (1990), it is the construction of hierarchically ordered, increasingly complex levels of understanding and the attempt to describe the logical relations among these categories that most clearly distinguishes phenomenography from other qualitative research methods.

The six response categories in figure 6.3 describe a developmental understanding of atomic structure (Renström, Andersson, and Marton, 1990). "The six conceptions of matter should not be seen as a set of one correct and five erroneous conceptions. At each level some new insights are added that cumulate to the kind of understanding aimed at. All these various understandings are packed into 'the correct' understanding of matter. In a way, we can see our investigation as laying free or making visible the various tacit, taken-for-granted layers of the (scientific) understanding of matter" (p. 568).

These authors go further in unpacking the developmental understanding of matter and explain, using figure 6.3, how the "scientific" understanding of atomic structure (Level F) builds through the hierarchy of levels. In Level B, and above, there is an understanding that substances exist in different forms and can

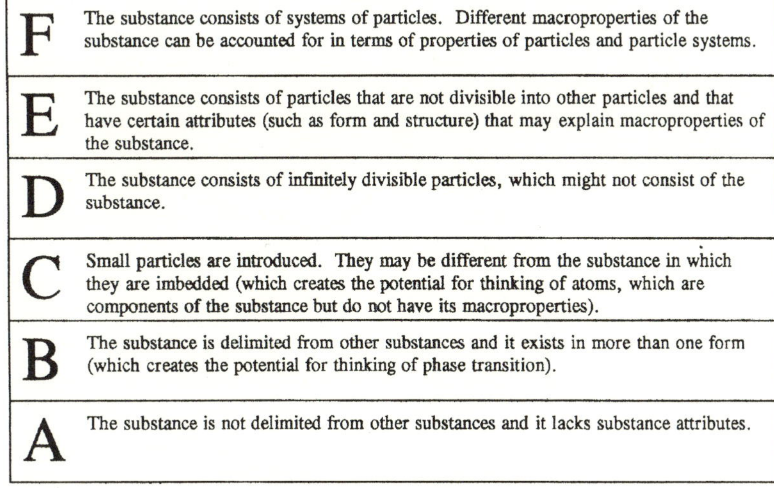

Figure 6.3. Levels in understanding of atomic structure.

change from one *state* to another (e.g., solid to liquid). In Level C, and above, there is a recognition of the existence of *atoms* (although in Level C, itself, these are thought of as particles embedded in the substance). In Level D, and above, there is a recognition that substances themselves consist of *particles* (although in Level D these are seen as infinitely divisible). In Level E, and above, substances are conceptualized as consisting of particles that are not infinitely divisible and that have *attributes*. And in Level F, there is a focus on *systems of particles* in terms of which the macroproperties of a substance can be understood.


## INFORMATION GATHERING

New views of student learning demand information gathering procedures that extend beyond the traditional standardized multiple-choice tests. During the last decade, work on these procedures has been called *authentic, alternative,* or *performance assessment.* The key features of such procedures have been described by Aschbacher (1991, p. 276) as follows:

1. Students perform, create, produce, or do something that requires higher level thinking or problem solving skills (not just one right answer).
2. Assessment tasks are meaningful, challenging, engaging, instructional activities.
3. Tasks are set in a real-world context or a close simulation.
4. Process and conative behavior are often assessed in place of, or as well as, product.
5. Criteria and standards for performance are public and known in advance.

Many of these features are not new (Stiggins, 1991). Forty years ago, Lindquist (1951, p. 152; also quoted in Linn, Baker, & Dunbar, 1991) wrote: "It should always be the fundamental goal of the achievement test constructor to make elements of his test series as nearly equivalent, or as much like, the elements of the criterion series as consequences of efficiency, comparability, economy, and expediency will permit." It is probably fair to say, however, that in the interim years concerns with "efficiency, comparability, economy, and expediency" have predominated. Multiple-choice tests have been advocated widely because they possess these characters. It is time to pay more attention to tasks that are valued because of their close alignment to the criteria of greater instructional importance.

The alternative assessment movement reminds us that there are many information-gathering formats. To facilitate discussion of these formats, we use the ideas shown in the "control chart" in figure 6.4. It does not describe all assessment types but, rather, assists us in describing several aspects of assessment that are relevant to this chapter. In the figure, the vertical dimension is used to indicate variation over the specification of assessment tasks. At the "high" end of this dimension we have assessment undertaken using externally set tasks that will only allow students to respond in a prescribed set of ways. Standardized multiple-choice tests are an example, whereas short-answer items are not quite at this extreme because students may respond in ways that are not predefined. The "low" end of this dimension is characterized by a complete lack of task or response specification. Teachers' holistic impressions of their students belong at this end of the task-control dimension. Lying between these two extremes are information gathering approaches, such as teacher-developed tests, and tasks that are adapted from central guidelines for local conditions. The horizontal dimension indicates control over judgment. The extremes are typified by machine scorability, at the "high" end, and unguided holistic judgments at the other end. Variations between these relate to the status of the judge and the degree of prescription provided by judgment protocols.

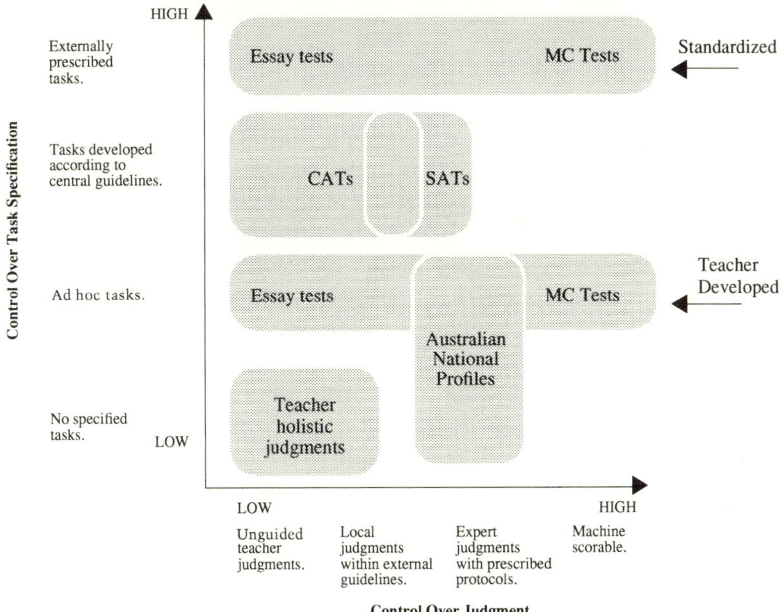


Figure 6.4. Control chart for assessment formats: Examples.

We have illustrated some examples of assessment formats that can be arrayed along these dimensions in figure 6.4. At the top of the figure are standardized tests of various sorts. Multiple-choice (MC) tests are represented on the righthand side because they can be machine scored, but essay tests can be judged in a variety of ways, so they occupy a broader range on the left. The same horizontal classification can be used for teacher-developed tests, but they appear lower on the task-specification dimension. In the bottom lefthand corner are holistic teacher judgments that may, for example, be made from memory without reference to specific tasks. The extension of the shaded region to the right allows teacher knowledge of general guidelines to be incorporated into judgments. Another region on the figure is exemplified by the Australia National Profiles (Australia Education Council, 1992) wherein teachers are provided with carefully prepared rating protocols that they use with ad hoc examples of student work or on the basis of their accumulated experience with students. Curriculum-embedded tasks, such as the UK Standardized Assessment Tasks ([SATs], Department of Education and Science, 1987b) and the Victoria Common Assessment Tasks ([CATs], Stephens, Money, & Proud, 1991) are externally specified project prompts that are interpreted locally to suit student needs and then scored by teachers. Within the CATs, control over the scoring varies from unguided teacher judgments to local teacher judgments within external guidelines. Typically, SATs involve a tighter control over judgment and have therefore been placed a little further to the right.

Assessments that are placed in different locations on the figure are often valued for different reasons. In figure 6.5, we indicate that assessments in the upper righthand corner are valued typically because they are perceived to have greater reliability; that is, they are composed of standardized tasks that are the same for all students, that can be scored using objective criteria, and that are congruent with existing psychometric models. Alternatively, assessments in the bottom lefthand corner are perceived typically to have greater instructional validity. That is, they are closer to the actual format and content of instruction, are based on the accumulated experience of teachers with their students, and allow adaptation to local conditions. It is desirable to have the positive features of both of these forms of assessment, but, as the figure illustrates, no single assessment format encompasses them.

The assessment net uses information obtained from a variety of locations on the figure; some information enhances validity and other information increases reliability. The new student assess-

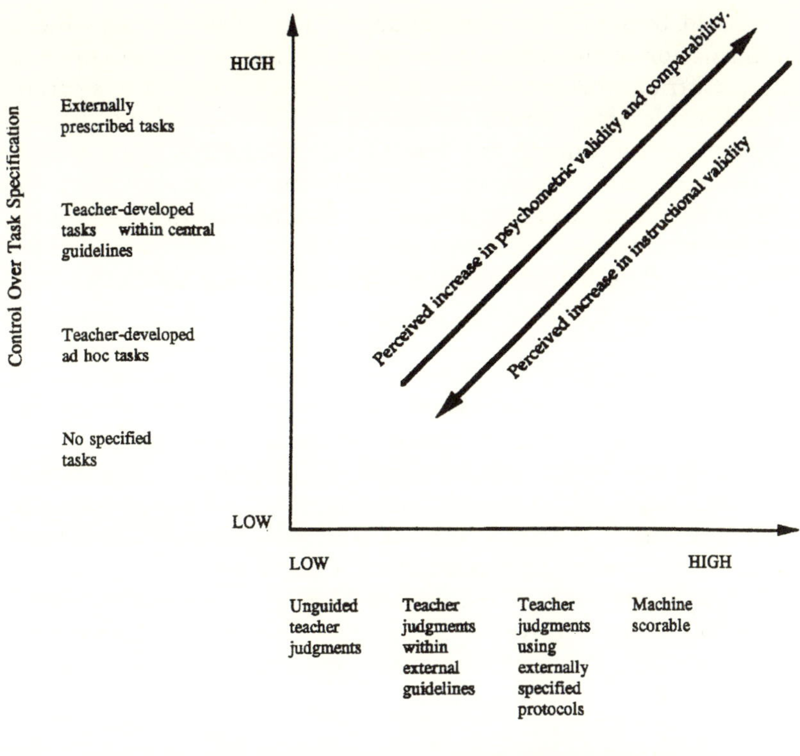


Figure 6.5. Control chart for assessment formats: Perceived advantages at extremes.

ment system being designed for Californian students (California Assessment Policy Committee, 1991) is an example of an assessment net. It is composed of three types of assessment activities.

*Structured on-demand assessments* include most forms of traditional examinations. These may range from 15-minute quizzes (of multiple-choice or open-ended format), to extended activities that can take up to three class periods, to a performance-assessment mode. Their distinguishing feature is that, although they derive from the framework in the same way as a student's regular instruction, they are organized in a more testlike fashion, with uniform tasks, with uniform administration conditions, and with no in-depth instructional activity occurring while they are taking place. The on-demand assessments could typically be scorable in a manner that involves little judgment on the part of the scorer, or could be scored by expert judges. This class of assessment information would reside at the top righthand corner in figure 6.4.

*Curriculum-embedded assessments* are to be seen as a part of their regular instruction by students. They would be chosen, however, from among the best of the alternative assessments, collected, tried out, and disseminated by teams of master teachers. They would typically be scored by the instructing teacher, although the results could go through certain types of adjustment for particular uses. This class of information would reside near the middle of figure 6.4.

*"Organic" portfolio assessments* include all materials and modes of assessment that a teacher or student decides should be included in a student's record of accomplishments. They can include a varied range of assessment formats and instructional topics. Teacher judgment on the relationship between these records and the levels in the frameworks are the major form of assessment information derived from the portfolios. This information finds its place in the bottom lefthand corner of figure 6.4.

Although each of the modes of assessment makes useful contributions to the overall assessment, what is needed is a way to integrate them and to ensure quality control.


## QUALITY CONTROL

A procedure is needed to coordinate the information (in the form of scores, ratings, or other data) that comes from the several forms of assessment. Procedures are also necessary (1) to examine the coherence of information gathered, (2) to map student performance, and (3) to describe the structural elements—items and raters—in terms of the strands or continua. Validity, reliability, bias, and equity studies must be carried out within the procedure. To meet these needs, we propose the use of generalized item-response models (sometimes called *item-response theory*). Generalized item-response models such as those described by Adams and Wilson (1992a), Kelderman (1989), Linacre (1989), and Thissen and Steinberg (1986) have now reached levels of development that make their application to many forms of alternative assessment feasible. The output from these models can be used for quality control and to obtain student and school locations on continua that may be interpreted both quantitatively and substantively.

We take an approach, based on Rasch-type models, because we need:

1. A latent continuous variable as an appropriate metaphor for many important educational variables;
2. The flexibility to use different "items," raters, etc., for different students if we are to reap the promised benefits of novel assessment modes;
3. A measurement approach that is self-checking—in this case, it is termed fit assessment;
4. A simple building block for coherent construction of complex structures;
5. A model that can be estimated efficiently from such basic observations as counts.

This last, (5), is of considerable importance because it corresponds to traditional educational practice for scoring tests and other instruments.

The next section may present a particular challenge because it supports some of its ideas using notation with which readers may be unfamiliar. Although we will not give full details on the statistical model, we will describe briefly some of its key elements to illustrate how it can meet the flexibility requirements that alternative assessments demand. Those educators who wish to gain more in-depth information on the model or apply the model in a school or district setting are referred to a detailed description of a unidimensional version of the model and a marginal maximum likelihood algorithm used to estimate its parameters in Adams and Wilson (1992a) and a multidimensional version in Wilson and Adams (1992a).

Suppose a test is composed of several items ($I$), where each item has a number of response categories ($K$). The observed response of any students to one of the items can be placed in one of the mutually exclusive categories represented by $K$. We have used the term *item* generically here. The items can, however, represent much more complex phenomena than the traditional multiple-choice test items we are used to (Wilson & Adams, 1992a). The items could represent an entire set of questions that relate to a common piece of stimulus material, for example, an item bundle or a testlet, and the response categories could be the response sets for the bundle. Or the items could represent a set of tasks that have been scored by a group of raters, where the rater-task pairs are considered as the item.

A statistical procedure allows scores to be assigned to a student's performance on each item and a vector developed that represents student abilities; it can be used to locate or map their performance on the strands or continua within the curriculum framework. Readers who are unfamiliar with matrix algebra may wish to pass through this section quickly and dwell on the example from Siegler that begins on page 000.

The vector $\mathbf{x}_{ni} = (\mathbf{x}_{ni2}, \mathbf{x}_{ni3}, \ldots, \mathbf{x}_{niK_{n-1}})'$ is used to denote the responses of person $n$ to item $i$, with a 1 placed in the category in which he or she responded, and a 0 elsewhere. Note that a response in the first category (which we are using as a reference category) is denoted by a string of zeroes. By collecting the item vectors together as $\mathbf{x}_n = (\mathbf{x}'_{n1}, \mathbf{x}'_{n2}, \ldots, \mathbf{x}'_{nK_{n-1}})$, we can formally write the probability of observing the response pattern as

$$f(\mathbf{x}_{n};\ \ \mathbf{A}, \mathbf{B}, \xi \mid \theta) = \frac{\exp \mathbf{x}'_n (\mathbf{B}\theta + \mathbf{A}\xi)}{\sum\limits_{z \in \Omega} \exp \mathbf{z}'(\mathbf{B}\theta + \mathbf{A}\xi)}, \qquad (6.1)$$

where $\mathbf{A}$ is a design matrix that describes how the elements of the assessments (e.g., raters and tasks) are combined to produce observations, $\xi = (\xi_1, \xi_2, \ldots, \xi_p)'$ is a vector of the parameters that describe those elements, $\mathbf{B}$ is a score matrix that allows scores to be assigned to each performance, and $\mathbf{u} = (\mathbf{u}_1, \mathbf{u}_2, \ldots, \mathbf{u}_d)'$ is a vector of student abilities, or locations on the framework continua. The summation in the denominator of (6.1) is over all possible response patterns and ensures that the probabilities sum to unity. The model is applied to particular circumstances by specification of the $\mathbf{A}$ and $\mathbf{B}$ matrices.

For example, consider the simplest unidimensional item-response model, the simple logistic model (SLM), otherwise known as the Rasch model (Rasch, 1980). In the usual parameterization of the SLM for a set of $I$ dichotomous items, there are $I$ item-difficulty parameters. A correct response is given a score of 1 and an incorrect response is given a score of 0. Taking a test with just three items, the appropriate choices of $\mathbf{A}$ and $\mathbf{B}$ are

$$\mathbf{A} = \begin{bmatrix} 1 & 0 & 0 \\ 0 & 1 & 0 \\ 0 & 0 & 1 \end{bmatrix} \quad \text{and} \quad \mathbf{B} = \begin{bmatrix} 1 \\ 1 \\ 1 \end{bmatrix}, \qquad (6.2)$$

where the three rows of $\mathbf{A}$ and $\mathbf{B}$ correspond to the three correct responses, the three columns of $\mathbf{A}$ correspond to the three difficulty parameters, one for each item, and the single column of $\mathbf{B}$ corresponds to the student location on the continuum.

If the $\mathbf{A}$ and $\mathbf{B}$ matrices given in (6.2) are substituted into (6.1), it can be verified that this is exactly the Rasch simple logistic model (see Adams & Wilson, 1992a). The estimated parameters that result from the application of the model would be a collection of item locations and person locations on a continuum.

More complicated item-response models may be expressed using equally straightforward matrices. For example, the partial credit model (Masters, 1982; Wilson, 1992b) is designed for assessment situations with multiple levels of achievement within each item. For an instrument with, say, three items and three categories in each, the categories scored 0, 1, 2, then the **A** and **B** matrices are

$$\mathbf{A} = \begin{bmatrix} 1 & 0 & 0 & 0 & 0 & 0 \\ 1 & 1 & 0 & 0 & 0 & 0 \\ 0 & 0 & 1 & 0 & 0 & 0 \\ 0 & 0 & 1 & 1 & 0 & 0 \\ 0 & 0 & 0 & 0 & 1 & 0 \\ 0 & 0 & 0 & 0 & 1 & 1 \end{bmatrix} \text{ and } \mathbf{B} = \begin{bmatrix} 1 \\ 2 \\ 1 \\ 2 \\ 1 \\ 2 \end{bmatrix}. \tag{6.3}$$

The matrices have six rows, two for each item—recall that the responses scored **0** do not appear in the matrix because they are used

---

17. Evaluation of conclusions
 4. Interpreting and critically evaluating results
11. Understanding, interpretation, and evaluation of mathematics used

15. Defining mathematical symbols used
 3. Analysing information
 6. Breadth or depth of investigation

10. Use of mathematical language, symbols and conventions
 7. Mathematical formulation or interpretation of problem situation or issue
16. Account of investigation and conclusions

 9. Level of mathematics used
 5. Working logically
14. Relating topic to theme

12. Accurate use of mathematics
 1. Identifying important information
18. Organisation of material

13. Clarity of aims of project
 8. Relevance of mathematics use
 2. Collecting appropriate information

Figure 6.6. Victoria projects criteria ordered by average rating.

as reference categories. The **A** matrix is a block diagonal matrix indicating that each item is modeled by a unique set of parameters. The **B** matrix contains the scores allocated to each of the responses, and it has one column, corresponding to a single ability dimension. For example, the ordering of the Victoria project criteria using this model is given in figure 6.6. Such an ordering may be used to add depth to the interpretation of the variable being assessed by the projects. A similar ordering may be used to construct a continuum for the phenomenography example (see Masters, 1992) and could also be found for the Vermont portfolios, given suitable data.

A slightly more complicated example is provided by the ordered partition model (Wilson, in press), which, in addition to the features of the partial credit model, also allows for categories within an item to have the same score. For an instrument with, say, two items and five categories in each, the categories were scored 0, 1, 2, 2, and 3, respectively, then the **A** and **B** matrices are

$$
\mathbf{A} = \begin{bmatrix} 1 & 0 & 0 & 0 & 0 & 0 & 0 & 0 \\ 0 & 1 & 0 & 0 & 0 & 0 & 0 & 0 \\ 0 & 0 & 1 & 0 & 0 & 0 & 0 & 0 \\ 0 & 0 & 0 & 1 & 0 & 0 & 0 & 0 \\ 0 & 0 & 0 & 0 & 1 & 0 & 0 & 0 \\ 0 & 0 & 0 & 0 & 0 & 1 & 0 & 0 \\ 0 & 0 & 0 & 0 & 0 & 0 & 1 & 0 \\ 0 & 0 & 0 & 0 & 0 & 0 & 0 & 1 \end{bmatrix} \text{ and } \mathbf{B} = \begin{bmatrix} 1 \\ 2 \\ 2 \\ 3 \\ 1 \\ 2 \\ 2 \\ 3 \end{bmatrix}. \tag{6.4}
$$

The matrices have eight rows, four for each item. The **A** matrix is an identity matrix indicating that each response to each item is modeled by a unique response parameter. The **B** matrix contains the scores allocated to each of the responses, and it has one column, corresponding to a single ability dimension.

As an example of a continuum constructed using the ordered partition model, consider the example from Siegler (1987) described earlier. In his data, there were sixty-eight students with complete data records, from grades kindergarten, 1, and 2. The problems ranged from those as easy as "4 + 1 = ?" to the more difficult "17 + 6 = ?". For illustrative purposes, I have chosen a subset of the original item set: (A) 12 + 2, (B) 14 + 1, (C) 3 + 14, (D) 1 + 14, (E) 17 + 4, (F) 16 + 6. These are in three pairs: The first pair is taken from Siegler's problem type 1, where the larger addend is first and the smaller addend is relatively small (i.e., in the range 1–3); the second pair is taken from his problem type 2, which is the same except that the larger addend is second; and the third pair is taken from his problem type 4, where the larger addend is first and the smaller addend is relatively larger (in the range 4–6), which means that the sum is also relatively larger. Because there are six items, the appropriate **A** and **B** matrices will have three times as many rows as those shown in (6.4), but will otherwise have the same structure.

Figure 6.7 is a map of the continuum that has been constructed from the calibrated item difficulties for ratings of student

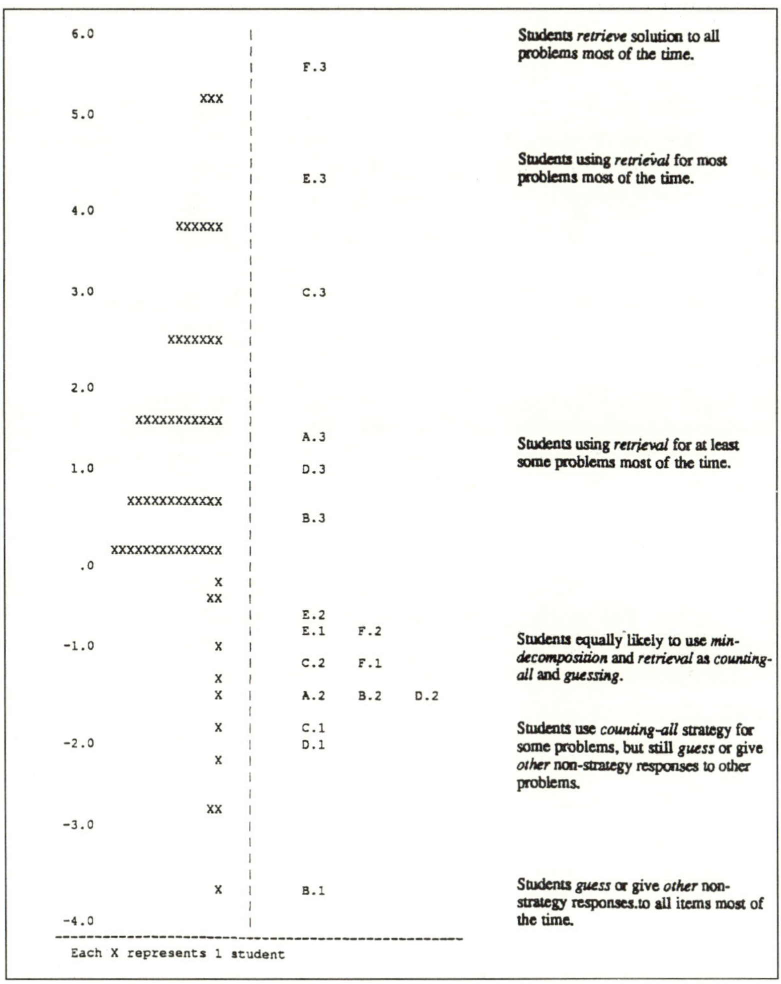

Figure 6.7. A continuum in addition strategy-use.

strategy-use on the six addition items. The map has a vertical scale—the numerical expression of the continuum—that represents increasing difficulty, and in the middle panel the difficulty thresholds for items are plotted. In this panel, we use the notation $X.n$ to indicate the difficulty of achieving level $n$ in item $X$. The left side of the figure indicates the distribution of student scores over the continuum. The map relies on the fact that the measurement model produces person-ability estimates and item-difficulty estimates that are expressed on a common scale. A certain amount of detail is lost in using this map (in fact, information is provided only about the points at which successive cumulative probabilities reach .5), but this is always the case with a summary. The thresholds can be interpreted as the crest of a wave of predominance of successive dichotomous segments of the set of levels. For example, $X.1$ is the estimated point at which levels 1, 2, and 3 become more likely than level 0 (for item $X$); $X.2$ is the estimated point at which levels 2 and 3 become more likely than levels 0 and 1; and $X.3$ is the estimated point at which level 4 becomes more likely than levels 0, 1, and 2.

In the righthand panel of the map are descriptions of increasing competence with respect to strategy-use—this is the substantive expression of the continuum. These descriptions allow a substantive interpretation of the numerical location that is estimated for each student by the measurement model. For example, a student at the position denoted by 2.0 on the numerical continuum would typically be expected to have a pattern of strategy-use like that described by the adjacent description on the substantive expression of the continuum; that is, he or she would be expected to use retrieval most of the time for half the items ($A$, $B$, and $D$), but less for the other three items. The student would not yet be expected to display the patterns indicated by the descriptions above 2.0, and would be expected to have previously displayed the understandings indicated by the descriptions below 2.0.

If we collected additional data from these students at a subsequent testing, we would obtain a second location for each student on the continuum. Hence, we can measure progress using the locations on the numerical continuum, and we can interpret it using the levels on the substantive continuum.

As a second illustration of the flexibility of the measurement model, consider a more complicated example that may be more typical of alternative assessment. Students are given two problem-solving tasks, and two judges place the students' performances into one of the four categories. Category one represents no strategy and is assigned a score of 0, categories two and three represent alternative but less sophisticated strategies and are both scored 1, whereas the fourth category represents a superior strategy and is scored 2.

A model that allows for an estimation of the difficulty of the tasks, as well as the relative harshness of the raters and places the students on a single continuum, is given by the following **A** and **B** matrices:

$$\mathbf{A} = \begin{bmatrix} 1 & 0 & 0 & 0 & 0 & 0 & 0 \\ 0 & 1 & 0 & 0 & 0 & 0 & 0 \\ 0 & 0 & 1 & 0 & 0 & 0 & 0 \\ 0 & 0 & 0 & 1 & 0 & 0 & 0 \\ 0 & 0 & 0 & 0 & 1 & 0 & 0 \\ 0 & 0 & 0 & 0 & 0 & 1 & 0 \\ 1 & 0 & 0 & 0 & 0 & 0 & 1 \\ 0 & 1 & 0 & 0 & 0 & 0 & 1 \\ 0 & 0 & 1 & 0 & 0 & 0 & 2 \\ 0 & 0 & 0 & 1 & 0 & 0 & 1 \\ 0 & 0 & 0 & 0 & 1 & 0 & 1 \\ 0 & 0 & 0 & 0 & 0 & 1 & 2 \end{bmatrix} \quad \text{and} \quad \mathbf{B} = \begin{bmatrix} 1 \\ 1 \\ 2 \\ 1 \\ 1 \\ 2 \\ 1 \\ 1 \\ 2 \\ 1 \\ 1 \\ 2 \end{bmatrix} \tag{6.5}$$

**A** and **B** have twelve rows, corresponding to the three possible nonzero scores for each of the four "items" (rater-task combinations). The first six rows are for the tasks rated by rater 1 while the last six rows are for the tasks rated by rater 2. The first six columns of **A** correspond to task parameters analogous to those in (6.3), and the last parameter is a rater-harshness parameter—in this instance, estimating how the harshness of rater 2 compares to that of rater 1. The rows of the **B** matrix are simply the item scores, and because we are again assuming a single continuum, **B** has only one column.

In the preceeding situation, we have modeled only variation in rater harshness. This is a simplistic view of how raters may vary. Raters may also vary in the way they use the response categories; some may have a tendency to give more extreme responses, whereas others may prefer the middle categories. This, and many other possibilities, could be modeled through different choices of A. The most general approach would involve the estimation of a separate set of item parameters for each rater.

In the case of multiple raters, maps like that illustrated in figure 6.7 could be constructed for each rater, or they could be constructed for the average rater. In a quality control context, the ideal would be to use this approach to help raters align their judgments. When this alignment process has resulted in a sufficiently common map for all raters, we would need only a single map. In the case of large numbers of raters, the model can be respecified under the assumption that the raters have been sampled from a population, and the model would estimate characteristics of the rater population, in particular, the degree of variation between raters.

In using this model as part of an assessment net, one would need to apply the procedure to mixed-item formats. The technique described generalizes quite readily to such situations and allows the specification of different weights for different formats. For example, a teacher's end-of-year rating would occur in the model as one item, and this would need to be weighted according to beliefs about the relative importance of the summary end-of-year rating compared to ratings on specific tasks.

Quality control information is also available in the assessment net. Standard techniques for assessing reliability, validity, fairness, and equity are available because of the measurement model's status as an item-response model (Hambleton, Swaminathan, Cook, Eignor, & Gifford, 1978; Lord, 1980; Wright & Masters, 1982).

## CONCLUSION

The methods just suggested are based on existing technologies in assessment and measurement. Some, such as frameworks and alternative assessment, can hardly be said to be new. Others, such as the complex measurement models, are quite new (although a computer program is now available to implement them [Adams & Wilson, 1992b]). They will, nevertheless, need adaptation to the tasks and conditions of specific circumstances. This will require considerable research and development before large-scale applications can be made.

The examples described here have been unidimensional. In some circumstances, such as the incorporation of different types of raters—for example, fixed and random—into the assessment net, it makes sense to use a multidimensional approach. Such situations and appropriate modifications to the model are described in Wilson (1992b).

## REFERENCES

Adams, R. J., & Wilson, M. (1992a). "A random coefficients multinomial logit: Generalizing Rasch models." Paper presented at the annual meeting of the AERA, San Francisco.

Adams, R. J., & Wilson, M. (1992b). RCML (Computer program). Berkeley: Graduate School of Education, University of California.

APA, AERA, & NCME. (1985). *Standards for educational and psychological tests.* Washington, DC: Authors.

Aschbacher, P. R. (1991). "Performance assessment: State activity, interest, and concerns." *Applied Measurement in Education* 4, no. 4:275–288.

Ashcraft, M. H. (1982). "The development of mental arithmetic: A chronometric approach." *Developmental Review* 2:213–236.

Australia Education Council. (1992). *Mathematics profiles, levels 1–6.* Melbourne, Australia: Curriculum Corporation.

California Assessment Policy Committee. (1991). *A new student assessment system for California schools.* Sacramento, CA: Author.

California State Department of Education. (1985). *Mathematics framework for California public schools, kindergarten through grade twelve.* Sacramento: Author.

Department of Education and Science. (1987a). *Education reform: The government's proposals for schools.* London: HMSO.

Department of Education and Science. (1987b). *National curriculum task group on assessment and testing: A report.* London: HMSO.

Hambleton, R. K., Swaminathan, H., Cook, L. L., Eignor, D. R., & Gifford, J. A. (1978). "Developments in latent trait theory: Models, technical issues and applications." *Review of Educational Research* 48:467–510.

Kelderman, H. (1989). "Loglinear multidimensional IRT models for polytomously scored items." Paper presented at the Fifth International Objective Measurement Workshop, Berkeley, CA.

Linacre, J. M. (1989). "Many faceted Rasch measurement." Doctoral dissertation, University of Chicago.

Lindquist, E. F. (1951). "Preliminary considerations in objective test construction." In E. F. Lindquist (ed.), *Educational measurement,* pp. 119–184. Washington, DC: American Council on Education.

Linn, R. L., Baker, E. L., & Dunbar, S. B. (1991). "Complex, performance-based assessment: Expectations and validation criteria." *Educational Researcher* 20, no. 8:15–21.

Lord, F. M. (1980). *Applications of item response theory to practical testing problems.* Hillsdale, NJ: Lawrence Erlbaum Associates.

Lybeck, L. (1981). *Archimedes in the classroom: A narrative on the didactics of subject matter.* Gothenberg: Acta Universitatis Gothobergensis.

Marton, F. (1981). "Phenomenography—Describing conceptions of the world around us." *Instructional Science* 10:177–200.

———. (1986). "Phenomenography—A research approach to investigating different understandings of reality. *Journal of Thought* 21, no. 3:29–49.

———. (1988). "Phenomenography—Exploring different conceptions of reality." In D. Fetterman (ed.), *Qualitative approaches to evaluation in education,* pp. 176–205. New York: Praeger Publishing.

Masters, G. N. (1982). "A Rasch model for partial credit scoring." *Psychometrika* 47:149–174.

———. (1992). *The measurement of conceptual understanding.* Hawthorn, Australia: ACER.

———, Adams, R. A., & Wilson, M. (1990). "Charting of student progress." In T. Husen & T. N. Postlethwaite (eds.), *International Encyclopedia of Education: Research and Studies,* supplementary vol. 2, pp. 628–634. Oxford: Pergamon Press.

———, & Mislevy, R. (1992). "New views of student learning: Implications for educational measurement." In N. Frederickson, R. J. Mislevy, & I. I. Bejar (eds). *Test theory for a new generation of tests.* Hillsdale, NJ: Lawrence Erlbaum Associates.

Ministry of Education, Western Australia. (1991). *First steps 1992.* Perth, Western Australia: Author.

Neuman, D. (1987). *The origin of arithmetic skills: A phenomenological approach.* Gothenberg: Acta Universitatis Gothobergensis.

Ramsden, P. (1990). "Phenomenographic research and the measurement of understanding: An investigation of students' conceptions of speed, distance and time." *International Journal of Educational Research* 13:1990.

Rasch, G. (1980). *Probabilistic models for some intelligence and attainment tests.* Chicago: University of Chicago Press. (Originally published 1960)

Renström, L., Andersson, B., & Marton, F. (1990). "Students' conceptions of matter." *Journal of Educational Psychology,* 82, no. 3:555–569.

Shepard, L. A. (1991). "Psychometricians' beliefs about learning." *Educational Researcher* 20, no. 6:2–16.

Siegler, R. S. (1987). "The perils of averaging data over strategies: An example from children's addition." *Journal of Experimental Psychology: General* 116:250–264.

Stephens, M., Money, R., & Proud, S. (1991). "Comprehensive assessment at senior secondary level in Victoria." Paper presented at the ACER conference on Assessment in the Mathematical Sciences, Geelong, Australia.

Stiggins, R. J. (1991). "Facing the challenges of a new era of educational assessment." *Applied Measurement in Education,* 4, no. 4:263–273.

———, & Plake, B. (eds.). (1991). Special issue: Performance assessment. *Applied Measurement in Education* 4, no. 4, whole issue.

Thissen, D., & Steinberg, L. (1986). "A taxonomy of item response models." *Psychometrika* 49:501–519.

Vermont Department of Education. (1991). *Looking beyond "the answer."* The report of Vermont's Mathematics Portfolio Assessment Program. Montpelier, VT: Author.

Victoria Curriculum and Assessment Board. (1990). *Mathematics study design*. Melbourne: Author.

Webb, N., & Romberg, T. A. (1992). "Implications of the NCTM *Standards for mathematics assessment*." In T. A. Romberg (ed.), *Mathematics assessment and evaluation: Imperatives for mathematics educators*. Albany: SUNY Press.

Wilson, M. (1992a). "The ordered partition model: An extension of the partial credit model. *Applied Psychological Measurement 16*, no. 3:309–325.

———. (1992b). "Measuring levels of mathematical understanding." In T. A. Romberg (ed.), *Mathematics assessment and evaluation: Imperatives for mathematics educators*. Albany: SUNY Press.

———. (1992c). "Measurement models for new forms of assessment in mathematics education." In M. Stephens & J.F. Izard (eds.), *Reshaping assessment practices: Assessment in the mathematical sciences under challenge*. Hawthorn, Victoria, Australia: ACER.

———. (1992d). "The ordered partition model: An extension of the partial credit model." *Applied Psychological Measurement 16*, no. 4:309–325.

———, & Adams, R. J. (1992a). "Rasch models for item bundles." Paper presented at the annual meeting of the AERA, San Francisco.

———, & Adams, R. J. (1992b). "Evaluating progress with alternative assessments: A model for Chapter 1." Invited address to the conference on Curriculum and Assessment Reform, Boulder, CO.

Wolf, D. P., Bixby, J., Glenn III, J., & Gardner, H. (1991). "To use their minds well: Investigating new forms of student assessment." *Review of Research in Education 17*:31–74.

Wright, B. D., & Masters, G. N. (1982). *Rating scale analysis: Rasch measurement*. Chicago: MESA Press.

## 7 ❖ Connecting Visions of Authentic Assessment to the Realities of Educational Practice

*M. Elizabeth Graue*

What a difference a few years make! Discussions about assessment in the 1980s focused primarily on the technical aspects of developing measures of achievement and the use of tests to leverage educational reform (see, for example, Airasian, 1988; Burnes & Lindner, 1985; Cohen, 1987; Mehrens & Kaminski, 1989). Conversations about testing (and testing, rather than the broader term *assessment*, was in fact the topic) centered on the concerns of policy makers and measurement specialists trying to refine tests as tools for reporting achievement for accountability. Testing instruments were developed and used for primarily administrative purposes; the spotlight was on assessment that informed various publics about the efficacy of education in the United States.

Noticeably absent from discussions during this time were teachers and subject matter specialists. They did not possess the knowledge (defined in terms of technical principles of psychometrics or policy frameworks) or the power to enter the conversation; they were as often the target of activity as a party to development or implementation planning of new assessment practices. This gap was exacerbated in a system that imposed testing instruments on classroom settings for purposes other than instruction but that had instructional implications (Smith, 1991). Those tests, developed using psychometric theory and technique that only measurement experts could understand, were increasingly mysterious to teachers. Assessment theory and technique have developed very separately from curriculum theory and practice. Theoretically part of the very same educational process, testing and assessment were, in practice, the property of measurement experts. Instruction and curriculum were the province of teachers and content area that of specialists.

As dissatisfaction grew with the social effects of assessment practices and the information that test scores could provide, new ways of thinking about learning forced us to reevaluate the way we monitored the educational process. The dual forces of curriculum reform and the negative effects of test-driven "improvement" have widened the circle of participants in assessment discussions beyond the psychometricians and politicians to include content area specialists and, potentially, practitioners. In addition, the focus of these discussions has broadened beyond summative evaluations of achievement to include more formative assessments for instructional decision making. A ground swell of support for new ways of finding out about what students know and how they know it has shifted the discourse on assessment, broadening the issues seen as salient, the tasks judged to be appropriate, and the technology used in analysis.

The work presented in this book represents a new phase in the assessment discussion, an attempt to begin *a conversation* among those who have a stake in monitoring educational activities. A real attempt has been made here to broaden the vision of assessment from unidimensional instruments constructed to heighten the statistical properties of items. Here we have a vision of possibility that makes assessment the bridge for instructional activity, accountability, and teacher development.

This vision, which values assessment of process as well as of outcomes, is becoming an accepted value by much of the educational community. One of the key foundations of the new assessment movement is a strongly held commitment to the congruence of curricular beliefs and assessment activities. The idea of instructional alignment (Cohen, 1987), which was prominent in the 1980s, suggested that instruction should be directed by standards represented by test content. Increasingly, that idea has been turned around to suggest that assessment practice needs to match instructional philosophy and activity. The actions of the mathematics education community are far ahead of the rest of the educational field because much of the necessary groundwork has already been done. As a set of guiding principles for practice, the NCTM *Standards* (NCTM, 1989, 1991) are unique in their specificity and coherence. By beginning with issues of content and pedagogy and working toward assessment, the cart is finally placed behind the horse—we focus on building assessments that serve teaching and learning purposes and are congruent with our understanding of content structure and process.

In addition, this movement should highlight the fact that the formats we choose for instruction or assessment are just that; they are *choices that we make as professionals.* Rather than existing separately from practice as objective paths to truth, our assessment activities, from fill-in-the-bubble tests to interviews, are socially negotiated facts that are developed to provide information about constructs that "carry with them in the process of interpretation a variety of value connotations stemming from three main sources: the evaluative overtones of the construct rubrics themselves, the value connotations of the broader theories or nomological networks in which the constructs are embedded, and the valuative implications of the still broader ideologies about the nature of humanity and society that frame the construct theories" (Messick, 1981, p. 12). This idea is echoed in Cherryholmes's (1988) discussion of construct validity: "Construct validity and research discourses are shaped, as are other discourses, by beliefs and commitments, explicit ideologies, tacit world views, linguistic and cultural systems, politics and economics, and how power is arranged" (p. 428).

These choices represent a certain set of values and ideas about the way that content, students, and teachers interact in classroom settings. One of the choices made in this book, by all of the authors, represents some variant of social constructivist theory as a foundation for their discussions of assessment. In using social constructivism as a foundation for our practice, we shift to new ideas that rearrange those interactions in profound ways. In this context, we must be careful not to rely on forms that personify old models of content, learning, and teaching, even if they have track records and a foundation of evidence that appear to make their use defensible. Even that evidence has a history and philosophy that links it to very specific conceptions of the educational process. These choices bring with them new challenges and responsibilities. The work in this book represents some of our best efforts to struggle with the ramifications of the move to constructivist approaches to education and its embodiment in assessment.

Thus, there is significant coherence in the goals set out by the book's authors. Three general themes are prominent in their investigations of authentic assessment in the context of reform: The critique of assessment that is ongoing, the frameworks being developed to guide assessment, and the nature of assessment activity. I review these issues in the next section.

## CRITIQUE

The first theme, critique, lays out the problems inherent in our current assessment practices. The process of critique has been the most prominent one; the resulting message is that what we have been doing does not match our aims and our current knowledge base. All of the authors point to the implications of changing our instructional philosophy—they forcefully address the need to rethink assessment practices. This shift implies a new relationship between assessment and instruction. By its very nature, constructivist teaching requires a merging of the two activities.

This merging of assessment and instructional philosophy and practice has pragmatic implications for the form activities take and for the meaning that we place on results. De Lange points out that in measuring the process and outcomes of new forms of mathematics curriculum, we must structure our assessments using constructivist (or, as he suggests, realistic mathematics) theories. Using other types of evidence to measure effectiveness would be comparable to deciding whether someone should make the basketball team using criteria developed for the swim team. Both are sports, but the demands, processes, and issues are rather different. The development of new assessment theory, which is vital as we explore authentic assessment strategies, has been called for by Romberg and Wilson in these pages, as well as by others in the measurement community (Fredericksen & Collins, 1989; Linn, Baker, & Dunbar, 1991; Moss, 1992).

The most pointed discussion of this issue is put forward by Stake in his argument about the utility of standardized tests to inform our understanding of mathematics learning. His chapter reminds us of the strange bind we find ourselves in when we rely on standardized tests, as they as currently constructed, to portray the nature of educational activity in mathematics classrooms. The simplistic nature of testing instruments misrepresents the complexity of mathematics as a knowledge system and has led us to overestimate the utility of tests to inform educational practice. This points to the key idea that assessment instruments as tools are very good for some purposes and not so good for others. Keeping in mind appropriate uses for the tools we use is the only way we will get out of the testing morass that has characterized recent educational practice.


## FRAMEWORKS TO GUIDE ASSESSMENT

Coming out of the current atmosphere of critique are frameworks for assessment congruent with a reformed view of mathematics. The recognition that it is necessary to map out the theory, values, and practice of assessment in a way that goes beyond psychometric considerations is a new and welcome approach that highlights its situated nature. Romberg and Wilson, as well as Lajoie, set out to build a vision of authentic assessment: The authors are clear that new assessment practices in mathematics should be shaped jointly by the NCTM *Standards* and evolving knowledge about socially developed cognition. Sharing many basic values for the development of authentic assessment, the ideas presented in these chapters are not redundant but, instead, highlight one of the key concepts that this book contributes to the literature—that educational practice, from systems planning to the development of individual instructional and assessment tasks, should be infused with common theoretical threads. Only then can we hope to avoid the conflicts that have characterized educational change in the recent past.

Romberg and Wilson speak to the broad vision of assessment that comes out of commitment to the *Standards*, moving beyond statements regarding curriculum content and instructional activity and extending the ideas from this consensual agreement for practice into plans for an assessment system. Their discussion is characterized by sweeping strokes that call for integration across the diversity of activities in education. Lajoie's contribution can be seen in her explication of two components of the new theories of learning: (1) work in the areas of situated cognition and the more inclusive idea of social constructivism, and (2) her application of these ideas to build a framework for authentic assessment tasks.

## ASSESSMENT ACTIVITY

The third theme is one of activity: In what kinds of tasks should we engage students to find out what they know? This is the primary focus of Silver and Kenney's, de Lange's, and Wilson's chapters, each of which approaches the problem from very different vantage points. Working from a position most similar to current practice, Wilson focuses on the development of psychometric models of assessment that are amenable to formats more open than multiple choice. Wilson's theoretical model includes dimensions of performance and knowledge that are salient in a more complex view of learning. The flexibility built into this proposal is a great advance from the limited traditional models of measurement that collapse performance on items or a collection of items into a single category; including issues such as strategy choice, as well as type of answer given, makes a much richer picture.

A symbolic question needs to be asked in a context in which highly sophisticated theories are used in analysis: How does the value of these new ways of unraveling the meaning of performance on authentic tasks alienate those without technical expertise even further than we have alienated them in the past? For those who are not highly conversant with the mechanics of item response theory, this approach might appear counterintuitive when compared to the rich, local knowledge provided by authentic assessment. Wilson's model has great promise, particularly in its ability to blend a variety of types of information generated in the course of instruction. I bring up these questions to those in the measurement community to remind them that part of their job is to justify and explain theoretical models to all participants because the practice of assessment *has* been broadened beyond the technically proficient. If assessment is seen as communication, having a clear way to communicate its process and products is vital. Only then will people have the confidence to use the tools that we develop.

The chapter by de Lange contains an impressive assortment of assessment tasks that enhances the possibilities for the types of activities we can use to probe student understanding. Even reading through these tasks forces a rethinking of mathematics activity; it represents an active approach to the field that does not allow for simple questions or simple answers. These assessment tasks provide an example of activities that are simultaneously evaluative and instructional; it would be hard to imagine a student whose imagination could not be captured by this fresh new way to approach mathematical ideas. The author's discussion of issues in the development of assessment tasks provides us with new ways to think about how we construct activities. In posing the importance of levels in assessment, the role of context, necessary and sufficient information, and a variety of test formats, U.S. educators have the opportunity to think beyond simple open items or multiple-choice formats.

Importing these ideas into U.S. classrooms will require, however, further elaboration so that vaguely familiar ideas are not transformed in the old measurement model context. It would be very easy to collapse the idea of levels of assessment into either a simplified version of Bloom's taxonomy or in terms of item difficulty—and that is not the intent of the author. Incorporating novel ideas into existing schemas can be tricky, and in this case it appears doubly so.

Silver and Kenney orient their chapter in terms of assessment and instructional practice. They clearly differentiate between assessment for decision making outside the classroom and assessment that serves day-to-day teaching decisions. This is an important distinction, particularly at this point in social history. Our attention to tests and critiques of assessment practices has focused primarily on the role of external assessment in education. In reorienting us to the differences between external and internal forms of assessment and heightening the salience of internal assessment for valid instruction, we can begin to bridge the gap between instructional and assessment practice, weaving the two together so that they more adequately inform one another. Giving authority to nonquantitative strategies for tracking student growth provides the teacher with a wider range of alternatives for gathering information and underscores the importance of ongoing monitoring of classroom activity.

It is exciting to see extended discussions of nontraditional assessment tasks like those presented by de Lange and Silver and Kenney. They provide a glimpse of the possibilities available to teachers in their instructional practice. Such strategies, from innovative tasks to interviews, promise a wealth of rich, contextualized information. But as presented, they make the reader hungry for more—for a discussion of strategies beyond data collection. How do teachers manage this information, make sense of it, and communicate it? Wilson's model is available for large-scale use but would not be appropriate for the day-to-day analysis activities that teachers face. These new tasks and strategies will require intense analytical development, with attention to the fact that the philosophy of analysis must match the philosophy of instruction and data collection. Otherwise, it is quite likely that we will fall back on easy-to-develop single scores to conflate the elaborate information available in authentic assessment, all in the name of entering the results in the grade book or to justify judgments made to parents and administrators. Just as we have been advised that multiple strategies for assessment will be needed, it would be fair to guess that multiple strategies for analysis will be required as well.

## CONVERSATIONS STILL TO BE HAD

So what is missing in this book? Have we pushed the ideas of constructivist assessment as far as they can go? Several chapters remain unwritten, both in this text and in the discussions that have occurred about assessment in the rest of the educational community. As we have moved our discussions to new forms of assessment, we have focused on critiques of old models and suggestions of new strategies and tasks. What we have not explored too carefully are next steps that require us to give up many of the traditional views of testing: exploration of ways to understand and make use of the information gathered through authentic tasks, discussion of an assessment system that has very specific tasks and purposes, and examination of the social utility and impact these new forms will have on schooling and those who participate in it. These missing discussions suggest that we may have fallen prey to the very problem that de Lange warned us of—we have not stepped away from the old perspectives of assessment.

The absence of such discussions could be seen as a remnant of thinking from the perspective of the old models of measurement. Some of those working in the field are still clinging to notions that were developed in what Berlak (1992) calls *a psychometric paradigm* rather than moving completely into a *contextual paradigm* (p. 12). In old models, the standardized test as a tool included within it the prescription for action; the results of a test provided an assumed blueprint for what happened next. Low scores, premised on a ranking notion of achievement, indicated the need for remediation, usually through reteaching or placement into another instructional group. This is shown most clearly in Shepard's (1991) work, which indicated that testing directors in U.S. schools tended to work from a factory-oriented, behaviorist model of learning in their understanding of testing and achievement.

## LIFE AFTER TEST ADMINISTRATION

With new forms of assessment, the focus of the process does not end with the development of sophisticated items that tap multiple levels of understanding or with the development of technology to model traits that are seen to underlie performance on an instrument. In a very basic sense, the implementation of an assessment task is just the beginning—there is much to do *after* an instrument is given to a group of students. We have not pushed that idea as we have continued to rely on old notions of posttest activity in the form of scoring item performance. Although the discussion has expanded to include the use of both trained teachers and "experts" and involves a wider range of considerations than simply whether an answer is right or wrong, we are still not extending our discussions toward more interactive ways of understanding the nature of student activity. The focus of the conversation still veers toward standardized, end-point decisions regarding what a performance represents about some underlying trait.

If we are thinking about learning in the way that Romberg and Wilson suggest in their chapter, as an "image that is gradually brought into sharper focus as the learner makes connections, or perhaps like a mosaic, with specific bits of knowledge situated within some larger design that is continually being reorganized in an organic manner" (p. 5), then assessment should be seen as the lens used by both teachers and students to view the learner's increasing competence. Creating a view of assessment that is as dynamic as the process of learning requires us to broaden our vision of (1) what constitutes an assessment activity, (2) how that activity might be understood, and (3) how it might be used to improve instruction. The first step is what has been most clearly discussed in this book: A variety of tasks were proposed, with new ideas for scoring performance on items. Although an important activity in the total assessment enterprise, the suggestion of tasks and of criterion-related scoring rubrics does not take us to the end of the line.

I would argue that our next push in the field must be in a move away from singular attention to scoring items to *interpreting activity*. From this perspective, what we do is more like reading the meaning of what a student does rather than scoring what that student knows. It is a much more interpretive process, one that requires intimate knowledge of context, social relations, and the meaning of any particular act. Our visions of assessment must mirror our visions of learning—I would apply Romberg and Wilson's description of mathematics to the activity of assessment by replacing the word *mathematics* with the word *assessment:* Assessment is a set of rich, interconnected ideas. Assessment should be viewed as a dynamic, continually expanding field of human creation, a cultural product. Assessment is learning, directed by teachers and providing many ways of knowing for teachers, students, parents, and other publics.

This idea is hinted at by Romberg and Wilson in their discussion of the opportunity that authentic assessment provides for professional development; they give several examples of teachers making the professional judgments necessary to score open-ended forms of performance tasks, both in the United States and abroad. However, the argument is still framed in terms of scoring and emphasizes the need to "train" teachers so that they can see the underlying structure of student performance. Are there ways that we can open up the discussion?

One way to do this is to heed Moss's advice (1992) to explore alternative epistemological orientations for developing assessment and interpretation strategies. This approach has already been proposed by a number of scholars studying assessment alternatives (Berlak, 1992; Johnston, 1989) and has been attempted by a growing number of practitioners. A good example is the interpretive model for communicating about portfolios suggested by Moss et al. (1992). The framework they have developed encourages teachers and students to work together to make creative and intellectual choices reflecting local goals and interests. The teacher's interpretations of student work are the main component of the evaluation and are built through teacher narratives of student work, paired with samples of the work itself. This allows for grounded interpretations of student growth that can be reanalyzed by anyone who reads the case. It is distinctly different from an empiricist view of measurement; it is much more locally generated and qualitative. It relies on the professional knowledge of teachers and the relationships they have with their students.

## THE VARIED NATURE OF ASSESSMENT

Taking the approach just described honors the complexity inherent in any single performance or a group of performances: It contextualizes performance in the social setting in which it occurs. But to do this we need a clear picture of what the assessment enterprise is all about. One of the difficulties inherent in contemporary discussions of assessment is that we use the term *assessment* to refer to very different kinds of activities. On the one hand, it is used by Wilson to describe standardized, externally developed tasks and teacher judgments that are used to gain an understanding of a latent variable like student position along curricular continua. On the other hand, it is used by Silver and Kenney to describe strategies that teachers can use in the course of instructional practice to gather information about student learning. Somewhere in between fall de Lange's multilevel tasks.

As a whole the chapters in this book provide a glimpse of the elements of the assessment process in its totality, but the system, or the package, is not yet explicitly described. We have learned from recent history that when different forms of assessment encroach on one another, problems ensue. Explicitly mapping the questions to be answered, the decisions to be made, and the tools we develop to do the job is a vital first step as we move to integrated assessment programs. This is especially important in a transitional era such as the one we are going through today.

As forms of assessment are adapted to the changing needs of the educational system, it is vital to have a clear view of an *assessment system*, with all of its components, so that assessment activities can be balanced across needs. This requires explicit statements of the requirements of assessment paired with prescriptions for the implementation of appropriate strategies for generating information. Although all assessment is pointed to educational improvement, an *assessment system* is composed of elements developed to answer very different kinds of questions posed by very different publics. Questions of accountability to external audiences require the development of measures that examine broad visions of content and student performance within that content. The question of whether we are looking at national snapshots of skills, or state or district estimates of learning, will shape the nature of the tasks developed and the kinds of precision required in analysis. The practice of assessment in a local classroom requires very different levels of information, involving short-term judgment of students' developing expertise on highly specific educational content. We do not currently have the technical expertise to bridge these various assessment purposes (although Wilson's model is definitely a step in the right direction), and the political landscape of the schools makes attempting it questionable. Until we have confidence that the various assessment purposes are parallel, it is important to specify what level of assessment we are addressing in our conversations. It also requires that *all* forms of assessment be valued for their own specific purposes; that the authority for valid information be spread from the external and standardized instruments to include the appropriate use of strategies such as observation and interview.

## ARE THESE NEW FORMS OF ASSESSMENT USEFUL?

Linked to the issue of definition of the assessment enterprise is the need to develop ways to monitor the effectiveness and impact that new forms of assessment have on those involved in the educational system. Traditionally, the criteria against which assessment instruments and practices were measured included psychometric concepts like predictive validity, or bias, and economic factors like per-pupil cost. As we move to new forms of assessment, the criteria that we consider need to be congruent with the philosophy of the model employed, broadening to include assessment of impact on the school environment as a whole and to the social and institutional costs that go with new practices (Moss, 1992).

As we move through this era of reform in instructional and assessment practice, for example, we need to monitor the practical utility of the proposed assessment strategies, instruments, and analysis methods to ascertain whether they are doing the jobs we have set out for them. In particular, it is important to examine how new forms of standardized assessment provide information and their impact on classroom practice. The rationale for the development of formats like performance assessment is founded in part on the hope that it will provide more effective information to shape the practice of instruction (Wolf, Bixby, Glenn, & Gardner, 1991). Whether they can in fact accomplish this purpose is dependent on the subtle interaction of such factors as (1) our ability to develop the technical expertise to interpret performance on authentic tasks, (2) the practical utility of the strategy, and (3) the consequences connected with performance. We need to be mindful that nothing is inherently more useful, truthful, or real about new assessment tasks or strategies— their promise is directly related to the context in which they are used. The formats chosen represent a single piece of a larger picture. Rather the *practice of assessment*—the relationships formed, the interpretations made, the understandings generated, and the actions taken—defines the appropriateness of our assessment choices. The standards for validity developed in the psychometric paradigm have been expanded to include consideration of the consequences of assessment activity (Messick, 1989; Moss, 1992) and that standard still applies as we work to develop authentic assessments. Again, the utility of the tool is related to the task to which it is applied.

Calls for new forms of curriculum and assessment have come in part out of a concern about bias in the traditional approaches to teaching and testing. The distance between the life experiences of diverse learners and the activities in which we engage them is in theory bridged when we use real-life problems and assessment tasks in the classroom (Brown, Collins, & Duguid, 1989). One of the principles suggested by Lajoie for operationalizing authentic assessment in mathematics called for considerations of "racial or ethnic and cultural biases, gender issues, and aptitude biases" (p. 31). Measures of bias have typically focused on differential performance on tasks that sample outcomes of learning. As we move to more authentic forms of assessment, our view of bias must adapt to the philosophy and structure of the tasks that we propose and must be sensitive to issues of power in the implementation of assessment in the schools.

These assessment alternatives have implications for the relations among teachers, students, parents, administrators, and public audiences interested in education. They open up possibilities for participation, responsibility, and reflection; or they have the potential to increase alienation, capricious judgment, and lack of understanding. The authenticity of assessment relies on participant engagement in tasks and productive access to the information generated. Nothing is authentic about a task that systematically excludes people due to its structure. Ideas about bias should be broadened to include access to the process of assessment (i.e., How is student engagement facilitated by this strategy?) as well as productive understanding of the products it provides (i.e., How do parents understand the information presented in forms like portfolios, learning progress maps, or profiles?). We need to pay attention to the manner in which the tools of assessment invite or inhibit participation in their use, because this participation is the key to their utility. From this perspective, bias is a matter of equality of opportunity rather than equality of outcomes.

Connected to the need to monitor the use of new forms of assessment is the necessity of being mindful of the conditions under which teachers are asked to implement what amounts to radically different approaches to instruction and assessment. Essentially, teachers are being called upon to take responsibility and authority for assessment activities that were seen as technically beyond their expertise and that were, worse, tasks that the public did not trust them to do.

Traditional testing practices were developed from the factory model of education (Shepard, 1991), and the structure of most schools has retained that model in the form of age-graded groupings, grade-specific curricula, and standardized testing at almost every level. The new forms of assessment that have been proposed do not fit neatly into a compartmentalized or atomized system. If these new responsibilities are going to pay off for both teachers and students, then the school as an institution must adapt to the necessary change in order to facilitate the transformation. At the very least, any suggestion of reformed practice, whether it is instructional or assessment-oriented, needs to make parallel recommendations of structural change that make the reform possible. The NCTM *Standards* (1991) outline the need for teacher development in the area of curriculum, but, historically, assessment has been seen as very separate from instructional concerns. Teacher development in the area of assessment should be parallel or even embedded in teacher development in the area of instruction.

Given the labor-intensive nature of locally developed and relevant assessment practices, institutional changes will need to occur in teacher workload, staffing patterns, and economic investments in schools. For example, simple paper-and-pencil tests are attractive to many teachers because they are efficient ways to find out certain things about students. They do give us some information about outcomes, but other strategies tell us even more. Unfortunately, these alternatives to traditional tests take an immense amount of time and discipline to implement, both in terms of collecting data and in managing and communicating it (Gomez, Graue, & Bloch, 1991). Unless the structural constraints on assessment practice are taken seriously, the practices themselves may add to conditions that create inequality for teachers, students, and families (Apple, 1992). The burden of making this work has been placed quietly on teachers whose list of responsibilities gets longer every day and whose reputations have been shaped by questionable assessment policies in the past. Making reform work amounts to more than just changing people's minds about participating in good practice. The social context of the practice of assessment is just as important as the technical implications as we attempt to facilitate the development and implementation of authentic assessment.

## CONCLUSION

The work presented in this book can be seen as a second generation of discussions on authentic assessment. Taken as a whole, the authors' ideas move beyond suggesting the need for new approaches and describing isolated strategies for assessment in mathematics—they suggest an integrated assessment system that has a shared philosophical base. The coherence of the proposals provided here is one of the book's greatest strengths; in generating assessment ideas from within curricular content, the authors present models that avoid the fragmentation of instruction and assessment practice that has been such a problem in recent years.

In this transitional period, we are moving from atomistic views of teaching and testing to interactive visions of learning and assessment. In the process, we are shedding old ideas of what constitutes "real" information about student learning and reaching across disciplinary boundaries to construct new ways of knowing about the educational process. The key to making this work lies in continuing to explore the possibilities that exist, often in places where we might least expect to find them. In dismantling the boundaries that have held traditional measurement practice in place, we are more likely to come across ideas and strategies that will fit the needs of a reformed curriculum. New assessment theory and practice will need to be generated from within the frame of curriculum content as well as by the measurement community. Our best bet may be to form collaborations whose purpose is to develop strategies congruent with the spirit of the subject matter and within the limits of the environment in which that subject matter is taught and learned.

## REFERENCES


Airasian, P. W. (1988). "Measurement driven instruction: A closer look." *Educational Measurement: Issues and Practice* 7, no. 4:6–11.

Apple, M. W. (1992). "Do the *Standards* go far enough? Power, policy, and practice in mathematics education." *Journal of Research in Mathematics Education*, 25, no. 5:412–431.

Berlak, H. (1992). "The need for a new science of assessment." In H. Berlak, F. M. Newmann, E. Adams, D. A. Archbald, T. Burgess, J. Raven, & T. A. Romberg, *Towards a new science of educational testing and assessment*, Albany: State University of New York Press.

Brown, J. S., Collins, A., & Duguid, P. (1989). "Situated cognition and the culture of learning." *Educational Researcher* 18, no. 1:32–42.

Burnes, D. W., & Lindner, B. J. (1985). "Why the states must move quickly to assess excellence." *Educational Leadership* 43, no. 2:18–20.

Cherryholmes, C. H. (1988). "Construct validity and the discourses of research." *American Journal of Education* 96, no. 3:421–457.

Cohen, S. A. (1987). "Instructional alignment: Searching for a magic bullet." *Educational Researcher* 16, no. 8:16–20.

Frederiksen, J. R., & Collins, A. (1989). "A systems approach to educational testing." *Educational Researcher* 18, no. 9:27–32.

Gomez, M. L., Graue, M. E., & Bloch, M. N. (1991). "Reassessing portfolio assessment: Rhetoric and reality." *Language Arts* 68:620–628.

Johnston, P. H. (1989). "Constructive evaluation and the improvement of teaching and learning." *Teachers College Record* 90, no. 4:509–528.

Linn, R. L., Baker, E. L., & Dunbar, S. B. (1991). "Complex, performance-based assessment: Expectations and validation criteria." *Educational Researcher* 20, no. 8:5–21.

Mehrens, M. A., & Kaminski, J. (1989). "Methods for improving standardized tests scores: Fruitful, fruitless, or fraudulent." *Educational Measurement: Issues & Practice* 8, no. 1:14–22.

Messick, S. (1981). "Evidence and ethics in the evaluation of tests." *Educational Researcher* 10:9–20.

———. (1989). "Validity." In R. L. Linn (ed.), *Educational measurement*, 3rd ed. New York: Macmillan.

Moss, P. A. (1992). "Validity in educational measurement." *Review of Educational Research* 26, no. 3:229–258.

———, Beck, J. S., Ebbs, C., Matson, B., Muchmore, J., Steele, D., Taylor, C., & Herter, R. (1992). "Portfolios, accountability, and an interpretive approach to validity." *Educational Measurement: Issues and Practice* 11, no. 3:12–21.

National Council of Teachers of Mathematics (NCTM). (1989). *Curriculum and evaluation standards for school mathematics*. Reston, VA: Author.

———. (1991). *Professional standards for teaching mathematics*. Reston, VA: Author.

Shepard, L. A. (1991). "Psychometricians' beliefs about learning." *Educational Researcher* 20, no. 7:2–16.

Smith, M. L. (1991). "Put to the test: The effects of external testing on teachers." *Educational Researcher* 20, no. 5:8–11.

Wolf, D., Bixby, J., Glenn, J., & Gardner, H. (1991). "To use their minds well: Investigating new forms of student assessment." *Review of Research in Education* 17:31–74.

# CONTRIBUTORS

*Jan de Lange*
Freudenthal Institute
Utrecht, The Netherlands

*M. Elizabeth Graue*
University of Wisconsin
Madison

*Patricia Ann Kenney*
University of Pittsburgh

*Susanne P. Lajoie*
McGill University
Montreal

*Thomas A. Romberg*
University of Wisconsin
Madison

*Edward A. Silver*
University of Pittsburgh

*Robert Stake*
University of Illinois
Champaign

*Linda D. Wilson*
University of Delaware
Newark

*Mark Wilson*
University of California
Berkeley

# SUBJECT INDEX

alignment
    and authentic assessment, 30–1, 261
    and behaviorism, 4
    and curriculum reform, 7–8, 92–3
    definition of, 198
    difficulties of, 196–9, 214
    as simplification of curriculum,
      199
assessment feedback. *See* feedback
assessment levels
    higher level, 103–9
    lower level, 96–7
    middle level, 97–102
    reason for development of, 93–6
assessment nets
    California Framework as example
      of, 246–8
    discussion of, 236–59
Australian IMPACT Project
    as example of authentic assessment,
      31
authentic assessment. *See also*
    *Curriculum and Evaluation*
    *Standards for School*
    *Mathematics* (NCTM); reporting
    procedures; task development
    argument for development of, 2, 19,
      25, 168–9
    assumptions underlying, 2–4, 6,
      28–30
    and continuum maps, 253–56
    criteria for, 239–40, 251
    frameworks for, 19–37, 237–44, 248,
      263–4
    as instructional guidance, 38–86,
      265–6
    issues in, 1–18, 168–169, 269–274
    and learning theories, 11–12, 266–7
    principles for, 10–11, 30–31, 94–5
    quality control in, 237, 248–56

behaviorism
    and mathematics education, 4–5

British Committee of Inquiry into the
    Teaching of Mathematics in
    Schools
    goals of, 89
Bush, (President) George
    and national educational goals, 1,
      199, 201

California Achievement Test
    as example of standardized
      achievement tests, 41
California Assessment Program
    California Framework, 237–8, 246–8
    as example of authentic assessment,
      32
    and open-response tasks, 69, 70, 76
    and portfolios, 73
Chapter I programs
    and standardized achievement tests
      as criteria for placement in, 44–5
classroom tests. *See* teacher-made
    tests
Cognitively Guided Instruction
    as example of authentic assessment,
      32
cognitive psychology
    cognitive apprenticeship as model of
      instruction, 26–7
    and theories of learning, 5–7, 30, 75
College Board Advanced Placement tests
    and open-response tasks, 69
communication skills
    and California Assessment Program,
      32
    and classroom tests, 68, 145–7
    criteria for evaluation of, 70
    and *Curriculum and Evaluation*
      *Standards for School*
      *Mathematics* (NCTM), 9, 21, 25,
      59–60
    development of, in mathematics
      education, 23–4
    and equity issues, 24

computers
  effect of, on mathematics education,
    87–8, 92
  use of, in mathematics education,
    20, 22–3, 28–9
conjecturing
  as mathematics skill, 4, 20
connected curriculum
  use of, in mathematics education,
    24–5
Connecticut Common Core of
    Learning Project
  as example of authentic assessment,
    32
constructivism
  assumptions of, 5–6, 91
  as foundation for authentic
    assessment, 10–12, 262, 263, 264
  as instructional guidance, 91–3
  as result of changes in mathematics, 87
  and testing, 6–7
content inventories. *See also*
    curriculum; problem-solving
    skills; real-world activities;
    task development
  organization of, 178–9, 180–2, 194–5,
    223n
  and teachers, 182, 183–4, 214, 220n
core curriculum
  and decentralization of school
    management, 204
criterion-referencing. *See also* district/
    state-level tests
  emphasis of, 223n, 224n
curriculum. *See also* content
    inventories; public accountability;
    task development
  and alignment with testing, 41–3,
    46, 47, 196–9, 214
  change in, 92, 167, 201, 211–13
  and curriculum coordinators, 184
  effect of testing on, 213
  and International Assessment of
    Educational Progress (IAEP), 46,
    47
  and National Assessment of
    Educational Progress (NAEP), 49
  and national education reform, 7–8,
    38, 90
  and teachers, 199, 220n
*Curriculum and Evaluation Standards
    for School Mathematics* (NCTM)

  and authentic tasks, 25, 30–1, 76
  and balanced assessment, 162–3
  and development of standards, 1–2,
    7–9
  general assumptions of, 4, 7–8, 10–
    11, 21
  and goals of mathematics education,
    89–90
  and instructional guidance, 26, 58–
    60, 68, 72, 220n
  and learning theories, 5–6, 27, 264
  and National Assessment of
    Educational Progress(NAEP), 54
  and questioning as assessment tool,
    62
  and writing as assessment tool, 63

Department of Labor
  and criteria for high school
    graduation, 20
district/state-level tests
  changes in, 76
  as external assessment, 45–8
  as instructional guidance, 39

equity issues
  in assessment, 31, 34, 248, 271–2
  and item-response theory, 256
  in task development, 121–4
evaluation criteria
  in QUASAR assessment, 69–70
  in Vermont portfolios, 78
exploration
  as mathematics skill, 4
external assessment. *See also*
    standardized achievement tests;
    testing
  and classroom instruction, 40–56,
    56, 75–6, 265–6
  limitations of, 54–5, 75
  national/international surveys, 38,
    40, 48–9, 52–4
  reform of, 76, 270
  and scoring, 13
  state/district-level tests, 39

feedback
  adaptive (scaffolding), 26–7
  and computers, 23, 28–9
  and observation, 61

*First Steps* project (Western Australia)
  as example of assessment
    framework, 237
Freudenthal Institute
  and construction of tasks, 91, 156

Goals 2000
  and mathematics achievement, 48
graphs
  as assessment tool, 22–3, 53, 153–6
group activities
  and communication skills, 24, 25,
    58, 59
  and curriculum, 39
  design of, 28, 31, 32, 160, 164–6
  and problem-solving skills, 27

Hewet Mathematics Project (The
    Netherlands)
  as alternative assessment project, 33
homework
  as assessment tool, 71–2

International Assessment of
    Educational Progress (IAEP)
  and proficiency levels, 49, 50–1
  reason for existence, 202
interrater agreement
  and external assessment, 71, 255–6
Iowa Test of Basic Skills
  as example of standardized
    achievement tests, 41
item-response theory (IRT)
  as psychometric model, 6, 265
  and quality control of assessment,
    248–56

journals
  use of, in mathematics education,
    31, 63–65, 74

learning theories
  and mathematics education, 4–6, 15,
    23, 88, 91–2

mathematical hierarchies
  arguments against, 3

and phenomenographical analysis,
    242–4
and standardized achievement tests,
    8
mathematics ability
  difficulty of testing, 181–2, 221n
mathematics achievement
  artificiality of, as construct, 179–81,
    220n
  complexity of assessment, 181–3,
    236
  as instructional guidance, 191
Mathematics Contest in Modeling
  example from, 127
mathematics education. *See also*
    public accountability
  and conceptual fields (Vergnaud), 15
  definitions of, 174, 175
  dissociation of topics, 181–3, 221n
  mathematical categories as teaching
    strategy, 179
  mathematical literacy, 4, 20
  sample topics in, 176
multiple-choice formats. *See also*
    standardized achievement tests
  alternatives to, 31–3, 265
  construction of, 133–7
  criticisms of, 41, 43, 55, 68, 264
  and group activities, 164–5
  and machine scoring, 12, 245–6
  use of, in mathematics testing, 6,
    45, 54, 244–6, 247

National Assessment of Educational
    Progress (NAEP)
  and comparisons of groups, 202
  and *Curriculum and Evaluation
    Standards for School
    Mathematics* (NCTM), 54
  reports of, 49, 52–4
National Council of Measurement in
    Education (NCME)
  and testing mandates, 204
National Curriculum Profiles
    (Australia)
  as example of assessment
    framework, 237, 246
National Curriculum strands (United
    Kingdom)
  as example of assessment
    framework, 237

national education reform. *See also*
    content inventories; curriculum
  and decentralization of school
    management, 204
  goals of, 1, 199, 201, 212
  influence of district/state-level tests
    on, 47
  involvement of teachers, 200–1, 204,
    223n, 225n, 272–3
  and public accountability, 87–8, 200,
    202, 203–4, 260
  and public discussion of national
    examination system, 38, 48, 79n,
    199, 204
  reasons for, 87–8
National Governors Association
  and national educational goals, 1
norm-referencing. *See also*
    standardized achievement tests
  use of, in testing, 223n, 224n

observation
  as assessment tool, 39, 40, 60–1
open-response tasks
  as assessment tool, 40, 66–8, 242
  and California Assessment Program,
    32
  and classroom tests, 68–9
  construction of, 137–40
  and Hewet Mathematics Project
    (The Netherlands), 33
  in portfolios, 74
  scoring of, 11, 12

paper-and-pencil tests
  alternatives to, 31–8
  as assessment tool, 28, 39–40, 56
performance assessment. *See also*
    authentic assessment
  complexity of, 236–7
  formats of, 244–8
phenomenography
  and methods of assessment, 242–4,
    252
portfolios, use of
  in California Framework, 248
  criteria for, 239
  as format for assessment, 72–4, 76,
    160–2, 269

  in Vermont pilot project, 31–2, 73–4,
    78–9, 238, 252
problem-solving skills. *See also* task
    development
  in authentic assessment, 244
  in classroom tests, 68
  and Connecticut Common Core of
    Learning Project, 32
  in group activities, 27
  and Hewet Mathematics Project
    (The Netherlands), 33
  in homework, 71
  in mathematics curriculum, 21–3,
    25, 244
  as middle level assessment, 97–102
  and standardized achievement tests,
    41, 43
  and stategy-use models, 241–2
  and Vermont portfolios, 73
process-product paradigm
  and teacher-student interaction, 61–2
*Professional Standards of Teaching
    Mathematics* (NCTM)
  and instructional standards, 2
progress maps of learning
  as reporting model, 14–15
psychometric models
  and test theory, 6–7, 246, 260–1, 264
public accountability. *See also*
    national education reform
  effect on curriculum, 1, 211–13
  and reporting procedures, 14–16
  societal goals of mathematics
    education, 89–90

QUASAR project
  and assessment tasks, 69–70, 80n
QUASAR
  and Ford Foundation grant, 79n
questioning
  as assessment tool, 22, 40, 61–3

Rasch models
  use of, in quality control of
    assessment, 248–56
real-world activities. *See also* task
    development
  and equity issues, 121–4
  and ethics, 118–20

and problem-solving skills, 21, 28,
126–33
role of context in, 109–27, 168, 264
use of, in mathematics curriculum,
19–23, 25, 168, 244
reasoning skills
development of, in mathematics
education, 9, 21, 24, 25
and fragmented information tests,
156–9
and homework, 71
and standardized achievement tests, 43
reliability
of external assessment, 12–14, 246
and item-response theory, 256
of test items, 3
reporting procedures
and assessment, 1
models for, 14–16

scaffolding. *See* feedback
SCANS (Secretary's Commission on
Achieving Necessary Skills) test
and criteria for graduation from high
school, 20
scoring procedures
approaches to, 11, 12, 96, 167, 168
in Australia, 13–14
and authentic assessment, 1, 16, 30
and Connecticut Common Core of
Learning Project, 32
in The Netherlands, 12
Second International Mathematics
Study (SIMS)
and reports of results, 49
Secretary's Commission on Achieving
Necessary Skills. *See* SCANS test
self-assessment
and portfolios, 72–3
and systemic approaches, 29
and writing, 65–6
situated cognition
and application to mathematics
reform, 26–7, 264
and authentic activities, 20, 25
standardized achievement tests. *See
also* scoring procedures; student
placement/ranking; testing
content inventory of, 8, 193–6, 199,
211–12

ease of, for assessment, 96, 244–6
as evaluation of teaching, 202
influence of market on, 42–3
as instructional guidance, 40–5, 200
invalidity of, for measuring
achievement, 173–235, 263
as management tool, 174–5, 203–4
as measurement model, 3–4, 11–12,
173–5, 212
preparation for, 167
public view of, 191–2, 214, 237
valid interpretations of, 41, 173,
191–193, 196
Standardized Assessment Tasks
(United Kingdom)
as example of assessment format, 246
standards. *See Curriculum and
Evaluation Standards for School
Mathematics* (NCTM); reporting
procedures; task development
*Standards* (NCTM). *See Curriculum
and Evaluation Standards for
School Mathematics* (NCTM)
Stanford Achievement Tests
as example of standardized
achievement tests, 41
state level tests. *See* district/state-
level tests
student interviews
as assessment tool, 62–3, 242
design of, 63
student learning profiles
conceptual models of, 14–15
growth over time, 11–12
learning progress maps, 14–15
and use of portfolios, 72
student placement/ranking
and district/state level tests, 46–7
and quality control, 248
student ranking/placement
and standardized achievement tests,
44–5, 182, 193, 196, 213–14
systemic assessment
and Connecticut Common Core of
Learning Project, 32
definition of, 29

task development. *See also* problem-
solving skills; real-world activities
and California Framework, 247–8

task development *(continued)*
  complexity of, 121–4, 126–33, 168, 178–9, 265
  criteria for, 9–12, 244
  definition of authentic tasks, 19–20
  discussion of, 87–172
  and equity issues, 121–4
  formats for, 66–68, 133–63
  and Hewet Mathematics Project (The Netherlands), 33
  model building and pattern analysis, 21, 22
  superitems, discussion of, 33
  and test items, 2–3, 8–9, 52–3, 95–6
teacher-made tests (classroom tests)
  as assessment tools, 56–9, 68–9, 77, 203
teacher merit
  difficulty of measurement, 221n
teachers, role of. *See also* teaching
  as assessors, 12–14, 55–9, 213–14, 246, 248
  background in assessment, 13–14, 57, 168, 260, 268
  and curriculum, 39, 178, 183, 186, 198
  and definitions of assessment, 198, 225n
  as facilitators, 7, 62
  influence of external assessment on, 43–4, 46–7, 75–7, 200, 212–13
  interaction of, 174, 184
  perceptions of, on effects of testing on schools and instruction, 204–12, 215–19, 224–5n
  test results as evaluations of, 56, 191, 202, 214
  and use of journals, 63–4
  and use of questions, 61–2
teaching. *See also* teachers, role of
  complexity of, 175–9, 183–91, 221–22n, 223n
  conflict between official goals and actual practice, 196
  sample schematics of, 186–91
  standardization of, 185
technology. *See also* computers
  use of, in instruction, 21, 22, 28, 73
testing. *See also* external assessment; standardized achievement tests; task development; tests, formats for
  alignment of, 4, 75–6, 196–9, 202
  and comparison of groups, 202–3, 214, 224n
  content inventory of, 4, 8–9, 166–8, 173–4, 198–9
  design considerations of, 163–8, 193–6
  as education management, 203–4, 213
  as evaluation of teaching, 175, 214
  example teacher survey on effects of, 215–19
  and influence on teachers, 173–4, 203
  and instructional guidance, 43–4, 167, 192, 199–201
  as learning experience, 26–7, 29–30, 266–8
  and learning theories, 4–7
  and National Assessment of Educational Progress (NAEP), 52–3
  and placement/ranking of students, 12, 202
  purposes of, 94–6, 191–2
  teacher perceptions of, 204–12, 224n, 225n
tests, formats for
  combination of formats, 147–9
  project work, 159–60
  student production tests, 149–56
  timed tests, 164
textbooks
  problems with, 195, 196–8
  sample topics in, 176

validity
  of assessment, 246, 248, 270
  complexities of, 178, 205
  and item-response theory, 256
  teacher perceptions of, 204–6
  of test items, 2–3, 204–8
Vermont pilot project. *See* portfolios, use of
Victoria Curriculum and Assessment Board
  as model for external examination, 66, 77–9
  and projects as assessment, 238, 240–41
Victoria Common Assessment Tasks (CATs), 246, 251, 252

writing
  as assessment tool, 28, 31, 63–6,
    141–4
  and California Assessment Program,
    32
  and Hewet Mathematics Project
    (The Netherlands), 33
  in open-response tasks, 67–8
  in portfolios, 73–4
  as self–assessment, 65–6
WYTITYG/WYGIWICT (what you
    test is what you get/what you get
    is what I can teach)
  as assessment strategy, 76–7

# AUTHOR INDEX


Adams, E., 196, 226
Adams, R.A., 236, 258
Adams, R.J., 248, 249, 250, 256, 259
Adams, V.M., 68, 85
Airasian, P.W., 46, 56, 80, 82, 173, 226, 260, 274
Algina, J., 202, 229
Allen, R.A.B., 159, 169
American Educational Research Association (AERA), 173, 226, 237, 257
American Psychological Association (APA), 173, 226, 237, 257
Anacker, S., 6, 7, 17
Andersson, B., 243, 258
Aoki, T., 196, 226
Apple, M.W., 273, 274
Archbald, D., 9–10, 16, 195, 196, 226
Aschbacher, P.R., 244, 257
Ashcraft, M.H., 241, 257
Ashton-Warner, S., 220, 226
Australia Education Council, 237, 246, 257

Baker, E.L., 29, 36, 223, 226, 236, 244, 257, 263, 274
Ballew, H., 25, 34
Baron, J.B., 29, 32, 34
Barron, B., 25, 34
Beck, J.S., 269, 275
Belli, G., 173, 198, 228
Bennett, A., 62, 80
Bereiter, C., 221, 226
Berlak, H., 196, 226, 267, 269, 274
Bertin, J., 22, 34
Biggs, J., 220, 226
Bixby, J., 29, 37, 236, 259, 271, 275
Bloch, M.N., 273, 274
Bodin, A., 2, 16, 93, 169
Boertien, H., 125, 169
Borko, H., 55, 80
Bowman, N., 201, 226

Bransford, J., 25, 34
Bresler, L., 212, 234
Bridgeford, N.J., 56, 58, 86
British Columbia Ministry of Education, 60, 63, 65, 80
Britt, M., 68, 86
Broudy, H., 203, 226
Brown, C.A., 52, 80, 83
Brown, J.S., 26, 35, 271, 274
Brown, R.G., 80, 81
Bryant, P., 24, 34
Burgess, T., 196, 226
Burkhardt, H., 162–3, 169
Burnes, D.W., 260, 274
Burrill, G., 102, 129, 153–6, 162, 163, 169

Calfee, R., 212, 229
California Assessment Policy Committee, 247, 257
California Assessment Program, 32, 34
California Mathematics Council, 43, 81
California State Department of Education, 69, 81, 237–8, 257
Campbell, D., 226
Carey, D.A., 32, 35
Carey, N., 192, 234
Carl, I., 220, 226
Carlsen, W.S., 61, 81
Carpenter, T.P., 32, 34, 35, 52, 80, 83, 176, 233
Cazden, C.B., 62, 81
Chambers, B., 57, 58, 82
Chambers, D.L., 52, 66, 81
Charles, R., 60, 63, 65, 81
Chernak, R., 128, 169
Cherryholmes, C.H., 262, 274
Clarke, D., 31, 35, 66, 81
Cobb, P., 27, 37, 65, 84
Cockroft, W.H., 89, 95, 169
Cohen, E.G., 165, 169

Cohen, S.A., 260, 261, 274
Cole, N., 204, 226
Coley, R., 226
Collins, A., 26, 28, 29–30, 35, 72, 73,
    81, 263, 271, 274
Collis, K., 4, 17, 33, 35, 177, 212, 220,
    226, 227, 233
Connell, M.L., 63, 84
Connell, R., 183, 220, 227
Cook, L.L., 256, 257
Cooney, T. J., 49, 83, 88, 170, 202, 212,
    231
Corbett, D., 213, 235
Costello, D., 200, 227
Coulson, D.B., 202, 229
Cronbach, L., 173, 192, 224, 227
Crosswhite, F.J., 49, 83, 88, 170, 202,
    212, 231

Damrin, D., 224, 229
Danielsson, B., 232
Darling-Hammond, L., 43, 81, 204,
    208, 213, 223, 227
Davey, B., 32, 34
Davies, R., 221–2, 227
Davis, R., 222, 227
de Lange, J., 11, 12, 16, 33, 35, 87–172,
    212, 227, 263, 264, 265, 266, 267,
    269
Denny, T., 212, 234
Department of Education and Science,
    237, 246, 257
Department of Employment,
    Education and Training, 227
Dillon, J.T., 62, 81
Dorr-Bremme, D.W., 56, 81, 200, 229
Dossey, J.A., 49, 52, 65, 66, 81, 83, 88,
    170, 202, 212, 231
Doyle, W., 58, 81
Driver, R., 222, 227
Duguid, P., 271, 274
Duke, D., 221, 235
Dunbar, S.B., 29, 36, 236, 244, 257,
    263, 274

Easley, E., 179, 227
Easley, J., 179, 193, 212, 222, 227, 234
Ebbs, C., 269, 275
Edelman, M., 228

Eignor, D.R., 256, 257
Eiseley, L., 228
Eisner, E., 203, 228
Elliott, J., 221, 234
Ellwein, M., 211, 228
Ellwein, M.C., 57, 82
Elton, L.R.B., 58, 81
Eraut, M., 196, 228
Erickson, G., 222, 229
Erlwanger, S.H., 63, 82
Ernest, P., 3, 4, 5, 16

Fennema, E., 32, 34, 35
Fennessy, D., 56, 82
Fenstermacher, G., 222, 228
Fiske, D., 226
Fleming, M., 57, 58, 82
Floden, R., 173, 193, 198, 211, 228
Fong, G.T., 20, 35
Foreman, L., 62, 80
Forgione, P., 32, 34
Frechtling, J.A., 80, 82
Fredericksen, N., 26, 35
Frederiksen, J.R., 28, 29–30, 35, 263,
    274
Freeman, D., 173, 193, 198, 211, 228
Freudenthal, H., 91, 169
Furst, N., 61, 85

Gage, N., 185, 228
Gagné, R., 180, 228
Galbraith, 92
Galbraith, P.L., 91, 170
Garden, R.A., 136, 171
Gardener, H., 29, 37, 236, 259, 271,
    275
Gargiulo, S., 68, 86
Gifford, J.A., 256, 257
Giroux, H., 185, 228
Glaser, R., 7, 16, 75, 79, 82, 202, 223,
    228
Glass, G., 211, 228
Glenn, J., III, 29, 37, 236, 259, 271, 275
Goad, L., 196, 228
Godwin, R., 222, 228
Goertz, M., 226
Goffree, F., 109, 172
Goin, L., 25, 34
Goldstein, H., 202, 228

Gomez, M.L., 273, 274
Gong, B., 14, 17
Goodlad, J., 222, 228
Goslin, D., 200, 229
Graue, E., 224, 230, 260–75
Gravemeijer, K., 91, 97, 100, 103, 106, 111, 112, 170, 172
Green, J.L., 62, 82
Greeno, J., 25, 35, 229
Grimmett, P., 222, 229
Gronlund, N.E., 134, 141, 170
Grossman, R., 150, 170
Grumet, M., 222, 229
Gullickson, A.R., 56, 57, 82

Haertel, E., 56, 57, 82, 178, 193, 212, 220, 229
Halberstam, D., 223, 229
Hall, B., 205, 229
Hambleton, R., 182, 202, 229, 256, 257
Haney, W., 213, 229
Harmon, M.C., 76, 83
Harvey, J.G., 33, 35
Hasselbring, T., 25, 34
Hastings, T., 224, 229
HAVO, Mathematics A., 107, 113, 126, 140, 170
Hawkins, J., 28, 29, 30, 35
Henderson, K., 178, 229
Herman, J., 56, 81, 200, 229
Herter, R., 269, 275
Hill, S.A., 55, 82
Hively, W. II, 177, 229
Hoke, G., 212, 234
Hotvedt, M., 200, 230
House, E.R., 192, 230
Howson, G., 88
Hughes, P., 68, 86
Husen, T., 48, 82

Illinois State Board of Education, 97

Jaeger, R., 173, 204, 230
Jencks, S.M., 63, 84
Johnson, D., 179, 230
Johnson, R., 179, 230
Johnston, P.H., 269, 274
Johnston, W.B., 88, 170

Jonassen, D.H., 222, 230

Kaminski, J., 260, 274
Kelderman, H., 248, 257
Kellaghan, T., 56, 82, 200, 224, 231, 235
Kemmis, S., 205, 230
Kenney, P.A., 38–86, 264, 265, 266, 269
Khaketla, M., 43, 84
Kifer, E., 49, 83, 88, 170, 202, 212, 231
Kilpatrick, J., 42, 56, 57, 65, 68, 85
Kindt, M., 170
King, D.J., 224, 231
Klein, S., 78, 83
Komoski, K., 198, 230
Koretz, D.M., 78, 83, 200, 225, 230
Kouba, V.L., 52, 80, 83
Krantz, D. H., 20, 35
Kruglenski, H., 32, 34
Kuhs, T., 211, 228
Kulewicz, S., 25, 34
Kustiner, L.E., 128, 169

Lajoie, S., 7, 10, 16, 17, 19–37, 263, 264
Lamon, S., 14, 17
Lampert, M., 19, 20, 27, 35, 183, 230
Lane, S., 69–70, 86
Lapointe, A.E., 49, 50–1, 83, 88, 170
Lee, S.Y., 88, 172
Leinhardt, G., 43, 77, 83
LeMahieu, P., 43, 77, 83, 225, 230
Lesgold, A., 7, 16, 26, 29, 35
Lesh, R., 14, 17
Lester, F., 60, 63, 65, 81
Lieberman, A., 183, 225, 230
Linacre, J.M., 248, 257
Lindner, B.J., 260, 274
Lindquist, E.F., 244, 257
Lindquist, M.M., 52, 66, 80, 81, 83
Linn, R.L., 29, 36, 75, 80, 83, 173, 224, 230, 236, 244, 257, 263, 274
Littlefield, J., 25, 34
Lohman, D.F., 75, 86, 234
Lomax, R.G., 76, 83
Lord, F. M., 256, 257
Lorge, I., 192, 231
Lortie, D., 185, 231

Lundgren, U., 203, 222, 231
Lundin, S., 177, 229
Lybeck, L., 242, 257

Maassen, J., 170
Mabry, L., 212, 234
MacRury, K., 202, 231
Madaus, G.F., 56, 76, 80, 82, 83, 173, 204, 213, 224, 226, 229, 231
Mamona, J., 68, 86
Mandinach, E., 224, 234
Marton, F., 242–3, 257, 258
Marvin, C., 231
Masters, G.N., 236, 251, 252, 256, 258, 259
Mathematical Sciences Education Board (MSEB), 87–8, 95, 118, 161, 167, 170
Matson, B., 269, 275
Mattsson, H., 205, 231
MAVO, Mathematics A., 138, 170
Maxwell, G., 177, 229
McCaffrey, D., 78, 83
McDonnell, L., 192, 234
McKnight, C.C., 49, 83, 88, 170, 202, 212, 231
McLean, L., 193, 200, 231
McTaggart, R., 205, 230
Mead, N.A., 49, 50–51, 83, 88, 170
Meehl, P., 192, 227
Mehrens, M.A., 260, 274
Mehrens, W., 198, 231
Messick, S., 262, 271, 275
Midgley, M., 224, 231
Millman, J., 223, 231
Ministry of Education, Western Australia, 237, 258
Mislevy, R., 6, 7, 17, 236, 258
Mitzel, H.E., 61, 83
Money, R., 13, 17, 66–67, 77, 86, 246, 258
Morgan, K., 225, 234
Moss, P.A., 263, 268–9, 270, 271, 275
Mosteller, F., 20, 36
Mouseley, J., 232
Muchmore, J., 269, 275
Mullis, I.V.S., 52, 54, 65, 66, 81, 83
Mumme, J., 73, 84, 161, 162, 171
Murphy, S., 160–1, 171
Myrdal, S., 232

Nagy, P., 202, 231
National Assessment Governing Board, 54, 84
National Council of Teachers of Mathematics (NCTM), 1–2, 4, 5–6, 7, 8–9, 10–11, 16, 17, 21, 25, 26, 27, 30–31, 32, 36, 42, 44, 52, 54, 58, 59, 62, 63, 68, 72, 76, 84, 89–90, 121, 151, 162, 171, 189, 220, 225, 232, 261, 264, 272, 275
National Council on Measurement in Education (NCME), 173, 226, 237, 257
National Institute of Education, 75, 84
National Mathematics A-lympiade, 142, 158, 171
National Research Council, 42, 58, 84
Natriello, G., 58, 84
Neuman, D., 242, 258
Newman, S., 26, 35
Newmann, F., 9, 10, 16, 195, 196, 226
Nicholls, J.G., 65, 84
Nisbett, R.E., 20, 35
Niss, M., 171
Nitko, A.J., 29, 36, 75, 80, 84
Noe-Nygaard, E., 157, 171

Oakes, J., 192, 232, 234
Oaxaca, J., 88, 171
O'Daffer, P., 60, 63, 65, 81
Oldham, E.E., 136, 171
Oliver, D., 232
Owen, E.H., 52, 65, 66, 83

Packer, A.E., 88, 170
Pandey, T., 69, 73, 84
Payne, D.A., 80, 84
Pea, R.D., 171, 222, 232
Pearsol, J., 225, 234
Peck, D.M., 63, 84
Pereira-Mendoza, L., 21, 36
Perry, M., 88, 172
Peters, T., 224, 232
Peterson, P., 5, 17, 32, 35
Phillips, D.C., 93, 171
Phillips, G.W., 49, 50–1, 52, 65, 66, 83, 88, 170
Phillips, L., 128, 169
Phillipy, S., 205, 229

Piaget, J., 193, 232
Pipho, C., 202, 232
Plake, B., 236–7, 258
Pollak, H., 19, 36
Popham, J., 220, 223, 232
Popkewitz, T., 211, 232
Popper,K., 93, 171
Porter, A., 173, 193, 198, 211, 225, 228, 232
Post, T., 14, 17
Proppé, O., 232
Proud, S., 246, 258

Querelle, N., 135, 171

Rabehl, G., 173, 229
Rakow, E.A., 224, 231
Ramsden, P., 242, 243, 258
Rasch, G., 250, 258
Raths, J., 212, 234
Raven, J., 173, 196, 226, 232
Reed, S. K., 22, 36
Renström, L., 243, 258
Resnick, D.P., 80, 84, 232
Resnick, L., 5, 17, 19, 20, 27, 36, 80, 84, 88, 94, 162–3, 169, 171, 177, 185, 232
Reynolds, W.A., 88, 171
Rheinboldt, W.C., 88, 171
Rice, M., 232
Rindone, D., 32, 34
Robitaille, D., 233
Romberg, T., 1–18, 22, 30, 33, 35, 36, 37, 43, 84, 85, 91, 102, 129, 153–6, 169, 171, 175, 176, 179, 196, 198, 212, 225, 226, 233, 236, 259, 263, 264, 268
Roodhardt, 113
Rose, B., 63–64, 85
Rosenheck, M., 25, 36
Rosenshine, B., 61, 85, 185, 233
Ross, K., 205, 230
Rotberg, I.C., 48, 85
Ruesink, N., 106, 170
Runkel, P., 224, 229
Russell, H.H., 136, 171

Sacks, J., 200, 235
Salmon-Cox, L., 54, 56, 77, 85

Salomon, G., 29, 36
Sanders, N., 223, 230
Sarason, S., 222, 233
Sato, T., 222, 233
Saylor, G., 211, 233
Scheffler, I., 176, 179, 233
Schlesinger, B., 65, 68, 85
Schmidt, W., 173, 193, 198, 211, 228
Schoenfeld, A., 27, 36
Schön, D., 222, 234
School Mathematics Study Group, 222, 234
Schwartz, J., 33, 35
Schwille, J., 173, 198, 211, 228
Science Research Associates, 41–2, 85
Sension, D., 177, 229
Shaughnessy, J. M., 29, 36
Shavelson, R., 55, 65, 80, 85, 192, 234
Sheehan, N., 223, 234
Shell Centre for Mathematical Education, 68, 85
Shepard, L., 4, 6, 17, 43–4, 85, 204, 208, 213, 234, 237, 258, 267, 272, 275
Shohamy, E., 173, 234
Siegler, R.S., 241, 258
Silver, E.A., 38–86, 264, 265, 266, 269
Simons, H., 221, 234
Singer, J.D., 22, 37
Smith, G., 196, 228
Smith, M.A., 160–1, 171
Smith, M.L., 43, 86, 204, 208, 228, 234, 260, 275
Smyth, J., 205, 230
Snow, R.E., 28, 37, 75, 86, 224, 234
Soloway, E., 222, 232
Souviney, R., 68, 86
Stake, B., 225, 234
Stake, R., 173–235
Stecher, B., 78, 83
Steele, D., 269, 275
Steinberg, L., 248, 258
Stenmark, J.K., 44, 60, 62, 64, 65, 71, 73, 86
Stenzel, N., 212, 234
Stephens, M., 13, 17, 31, 35, 66–67, 77, 86, 246, 258
Stern, P., 65, 85
Stevenson, H.W., 88, 172
Stiggins, R.J., 56, 57, 58, 86, 221, 235, 236–7, 244, 258
Stigler, J.W., 88, 172

Stoeckinger, J., 176, 235
Streefland, L., 91, 97, 100, 104, 106, 150, 170, 172
Swafford, J.O., 49, 52, 80, 83, 88, 170, 202, 212, 231
Swaminathan, H., 256, 257
Swift, J., 21, 36

Tanner, M.A., 22, 37
Taylor, C., 269, 275
te Woerd, E., 106, 170
Theobald, P., 207, 211, 224, 235
Thissen, D., 248, 258
Thurston, W., 3, 17
Tittle, C., 173, 200, 230, 235
Torrance, H., 225, 235
Traub, R., 202, 231
Travers, K.J., 49, 83, 88, 134, 136, 170, 172, 202, 212, 231
Traxler, A., 173, 235
Treffers, A., 91, 109, 149, 172
Tres-Brevig, M. de P., 224, 235
Trowell, J., 68, 86
Tyson-Bernstein, H., 42, 86

United Nations Educational, Scientific, and Cultural Organization (UNESCO), 235

van den Brink, J., 104, 172
van den Heuvel-Panhuizen, M., 91, 97, 100, 103, 106, 111, 112, 170, 172
van der Kooij, H., 107, 144, 172
van der Ploeg, D., 106, 170
van Donselaar, G., 106, 172
van Manen, M., 222, 235
Vannatta, G., 176, 235
van Reeuwijk, M., 102, 129, 153–6, 169
Verhage, H.B., 151, 169
Verhoef, N.C., 170
Vermeulen, W., 106, 170

Vermont Department of Education, 73, 78, 86, 238, 258
Viator, K.T., 76, 83
Victoria Curriculum and Assessment Board, 99, 240, 259
Villeme, M., 205, 229
VWO, Mathematics A., 116, 172
Vygotsky, L. S., 26, 27, 37

Wainer, H., 22, 37
Waterman, R. Jr., 224, 232
Watson, J.M., 177, 227
Waywood, A., 31, 35
Webb, N., 236, 259
Weinzweig, A.I., 136, 171
West, M.M., 76, 83
Westbury, I., 134, 136, 172
Wheatley, G., 65, 84
Whetzel, D., 20, 37
Whitney, D.R., 80, 86
Wiley, D., 178, 220, 229
Willett, J.B., 22, 37
Williams, S.R., 43, 85, 225, 233
Willis, S., 235
Wilson, B., 213, 235
Wilson, J., 187, 235
Wilson, L., 1–18, 43, 84, 263, 264, 268
Wilson, M., 6, 13, 18, 236–59, 264, 265, 266, 269, 270
Wise, A.E., 43, 81, 213, 227
Wolf, D., 29, 37, 236, 259, 271, 275
Wood, T., 27, 37, 65, 84
Wright, B.D., 256, 259
W12–16, team, 100, 123, 172

Yackel, E., 27, 37, 65, 84
Yamamoto, K., 6, 7, 17
Yerushalmy, M., 33, 35

Zarinnia, E.A., 4, 15, 17, 22, 37, 43, 85, 212, 225, 233
Zawadowski, W., 109